Dimensions Math®
Workbook 3B

Authors and Reviewers

Jenny Kempe

Bill Jackson

Tricia Salerno

Allison Coates

Cassandra Turner

Singapore Math Inc.

Published by Singapore Math Inc.

19535 SW 129th Avenue
Tualatin, OR 97062
www.singaporemath.com

Dimensions Math® Workbook 3B
ISBN 978-1-947226-23-4

First published 2018
Reprinted 2019, 2020

Printed in China

Acknowledgments

Editing by the Singapore Math Inc. team.
Design and illustration by Cameron Wray with Carli Fronius.

Contents

Chapter	Exercise	Page

Chapter	Exercise	Page

Chapter	Exercise	Page

This workbook includes **Basics** (a review of basic concepts) and problems for **Practice**, for **Challenge**, and to **Check** understanding.

Chapter 8 Multiplying and Dividing with 6, 7, 8, and 9

Basics

1 (a) 6 × 6 is 6 more than ⬚ × 6.

6 × 6 = ⬚

(b) 9 × 6 is 6 less than ⬚ × 6.

9 × 6 = ⬚

2 (a) 8 × 6 = ⬚ + 24 = ⬚

⬚   4

(b) 7 × 6 = 30 + ⬚ = ⬚

5   ⬚

3 6 × 9 = 9 × ⬚ = ⬚

4 (a) ⬚ × 6 = 54   │   54 ÷ 6 = ⬚

(b) ⬚ × 6 = 36   │   36 ÷ 6 = ⬚

5 42 ÷ 6 = ⬚   │   42 + ⬚ = 45

45 ÷ 6 is ⬚ with a remainder of ⬚ .

Practice

6 Find the missing numbers.

$4 \times 6 = \boxed{}$ **P**	$18 \div 6 = \boxed{}$ **C**	$10 \times 6 = \boxed{}$ **T**
$54 \div 6 = \boxed{}$ **X**	$0 \times 6 = \boxed{}$ **E**	$7 \times 6 = \boxed{}$ **O**
$6 \times 2 = \boxed{}$ **I**	$42 \div 6 = \boxed{}$ **G**	$5 \times 6 = \boxed{}$ **H**
$6 \times 8 = \boxed{}$ **D**	$6 \times \boxed{} = 12$ **S**	$48 \div 6 = \boxed{}$ **K**
$6 \times 3 = \boxed{}$ **E**	$\boxed{} \div 6 = 6$ **C**	$9 \times 6 = \boxed{}$ **N**
$6 \times \boxed{} = 30$ **S**	$1 = \boxed{} \div 6$ **A**	$\boxed{} = 24 \div 6$ **,**

What are three animals to first be domesticated for food?

Write the letters that match the answers above to find out.

7	42	6	60	5	4	15	2	30	0	18	24	4

6	54	48	20	3	30	12	36	8	18	54	5	56

7 (a) $8 \times 6 = 6 + 6 + 6 +$ ⬚

(b) $9 \times 4 = 6 \times$ ⬚

8 (a) $23 \div 6$ is ⬚ with a remainder of ⬚.

(b) $59 \div 6$ is ⬚ with a remainder of ⬚.

(c) $50 \div 6$ is ⬚ with a remainder of ⬚.

9 Connor bought 6 packs of party favors.
Each pack came with 8 party favors.
He gave out 37 party favors.
How many party favors does he have left over?

Challenge

10 Aiden has $75.
He spent $28 on a pair of pants and $6 each on some shirts.
He has $5 left over.
How many shirts did he buy?

Basics

1 9 × 7 is [   ] less than 10 × 7.

9 × 7 = [   ]

2 (a) 7 × 7 = [   ] + 14 = [   ]

[   ]   2

(b) 6 × 7 = [   ] + 21 = [   ]

[   ]   3

(c) 8 × 7 = 40 + [   ] = [   ]

5   [   ]

3 7 × 8 = 8 × [   ] = [   ]

4 (a) [   ] × 7 = 63 | 63 ÷ 7 = [   ]

(b) [   ] × 7 = 28 | 28 ÷ 7 = [   ]

5 56 ÷ 7 = [   ] | 56 + [   ] = 62

62 ÷ 7 is [   ] with a remainder of [   ].

Practice

6 Find the missing numbers.

$3 \times 7 = \boxed{}$ **T**	$70 \div 7 = \boxed{}$ **A**	$6 \times 7 = \boxed{}$ **Y**
$49 \div \boxed{} = 7$ **C**	$63 \div 7 = \boxed{}$ **D**	$21 \div 7 = \boxed{}$ **N**
$9 \times 6 = \boxed{}$ **W**	$35 \div 7 = \boxed{}$ **I**	$5 \times 7 = \boxed{}$ **E**
$28 \div 7 = \boxed{}$ **A**	$\boxed{} \times 7 = 42$ **R**	$7 \times \boxed{} = 7$ **N**
$8 \times 7 = \boxed{}$ **O**	$56 \div 7 = \boxed{}$ **K**	$7 \times 7 = \boxed{}$ **L**
$9 = \boxed{} \div 7$ **S**	$7 \times 4 = \boxed{}$ **B**	$14 = \boxed{} \times 7$ **F**
$10 \times 7 = \boxed{}$	$8 \times 6 = \boxed{}$ **W**	$\boxed{} = 7 \times 0$ **O**

Joke: Why was the clock in the cafeteria always slow?
Write the letters that match the answers above to find out.

17	5	21	70	4	49	54	10	42	63	33

44	48	35	1	21	70	28	4	7	8	19

2	56	6	70	63	35	7	0	3	9	63

7 (a) $8 \times 7 = 7 + 7 + \boxed{}$

(b) $9 \times 7 = 7 \times \boxed{} + 35$

8 (a) $65 \div 7$ is $\boxed{}$ with a remainder of $\boxed{}$.

(b) $50 \div 7$ is $\boxed{}$ with a remainder of $\boxed{}$.

(c) $62 \div 7$ is $\boxed{}$ with a remainder of $\boxed{}$.

(d) $30 \div 7$ is $\boxed{}$ with a remainder of $\boxed{}$.

(e) $48 \div 7$ is $\boxed{}$ with a remainder of $\boxed{}$.

9 There are 35 days in _____ weeks.

10 There are _____ days in 9 weeks.

Challenge

11 A box of 10 gel pens costs $7.
Grace spent $56 on boxes of these pens.
Then she gave 7 pens to each of her friends.
She has 38 pens left.
How many friends did she give pens to?

Basics

1 Fill in the missing numbers or digits.

9 × 7 = ☐

60 × 7 = ☐

800 × 7 = ☐

869 × 7 = ☐

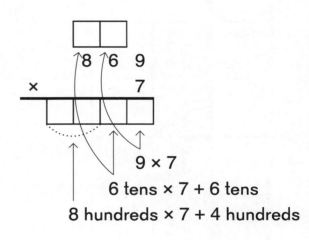

8 6 9
×   7
─────────
        ← 9 × 7
        ← 60 × 7
        ← 800 × 7

☐ ☐
⤴8 ⤴6 9
×   7
─────────
☐ ☐ ☐ ☐

9 × 7
6 tens × 7 + 6 tens
8 hundreds × 7 + 4 hundreds

2 6 × 8 = ☐

6 × 900 = ☐

6 × 908 = ☐

☐
9 0 8
×   6
─────────
☐ ☐ ☐ ☐

3 Use mental math to solve: 6 × 407 = ☐ .

Practice

4 Complete the cross-number puzzle using the clues below and on the next page.

Across	Down
G 80 × 6	**N** 6 × 21
N 306 × 6	**M** 107 × 6
Q 911 × 7	**A** 7 × 81

Across

720 × 7	6 × 36	7 × 13
A	**D**	**E**
89 × 7	435 × 7	521 × 6
I	**J**	**K**
6 × 161	7 × 76	47 × 6
L	**O**	**P**

Down

74 × 6	132 × 6	7 × 199
B	**C**	**F**
115 × 7	7 × 886	528 × 7
H	**I**	**J**

Basics

1 Divide 958 by 6.

6⟌9 5 8

9 hundreds ÷ 6 is _____ hundred with 3 hundreds left over.

35 tens ÷ 6 is _____ tens with 5 tens left over.

58 ones ÷ 6 is _____ ones with 4 ones left over.

958 ÷ 6 is [   ] with a remainder of [   ].

Check: [   ] × 6 + [   ] = 958

2 Divide 672 by 7.

7⟌6 7 2

67 tens ÷ 7 is _____ tens with 4 tens left over.

42 ones ÷ 7 is _____ ones with 0 ones left over.

672 ÷ 7 = [   ]

Check: [   ] × 7 = 672

Practice

3 Divide.

96 ÷ 7 **U**	84 ÷ 6 **R**	651 ÷ 7 **L**
756 ÷ 7 **A**	194 ÷ 6 **M**	508 ÷ 7 **L**
324 ÷ 6 **B**	242 ÷ 7 **E**	756 ÷ 6 **N**

Riddle: What goes up when rain comes down?

Write the letters that match the answers above to find out.

108	126	765	13 R 5	32 R 2	54	14	34 R 4	72 R 4	93	108

4 There are 123 red beads and 345 blue beads altogether.
They are distributed evenly into 6 bags.
How many beads are in each bag?

5 A couch costs 6 times as much as a chair.
The chair and couch together cost $168.
How much does the couch cost?

Challenge

6 A florist had 54 more roses than tulips.
After she sold 30 tulips, she had 7 times as many roses as tulips.
How many roses does she have?

Check

1 The product of the two numbers in the overlapping squares is the third number in each large square.
Fill in the missing numbers.

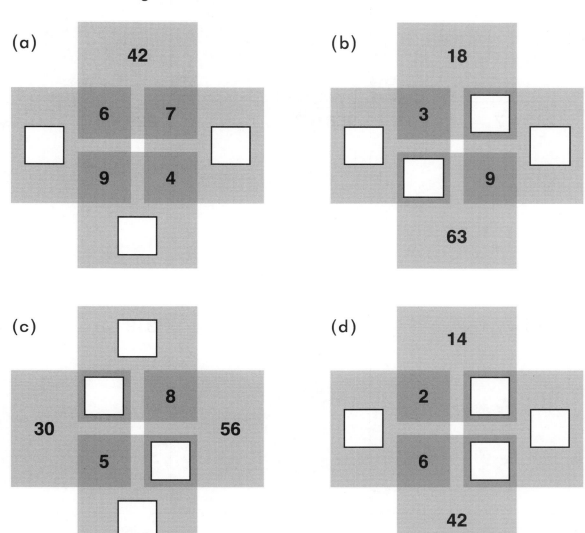

(a)

(b)

(c)

(d)

2 Multiply or divide.

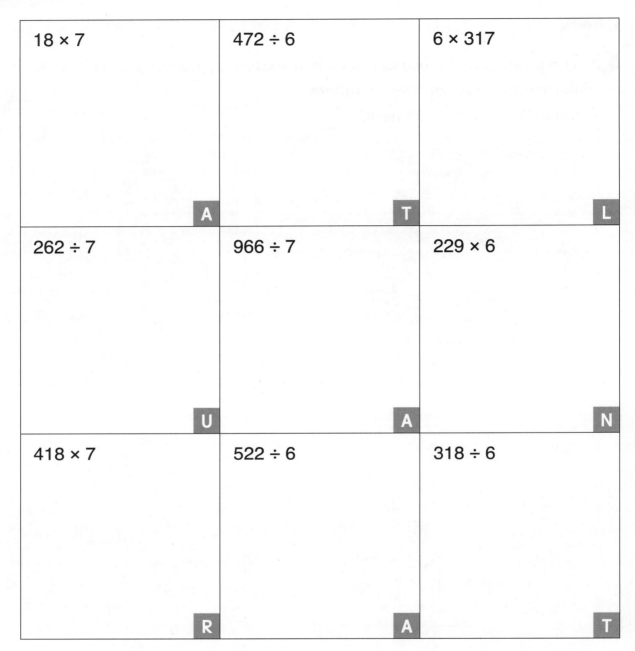

18 × 7	472 ÷ 6	6 × 317
A	T	L
262 ÷ 7	966 ÷ 7	229 × 6
U	A	N
418 × 7	522 ÷ 6	318 ÷ 6
R	A	T

What spider from the Amazon Basin is the largest type of spider in the world?
Write the letters that match the answers above to find out.

126	527	78 R 4	87	2,926	138	1,374	53	37 R 3	1,902	126

3 Write the missing digits.

(a)
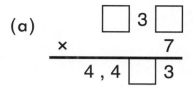

```
        □ 3 □
    ×       7
    4 , 4 □ 3
```

(b)
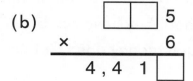

```
      □ □ 5
    ×     6
    4 , 4 1 □
```

(c)
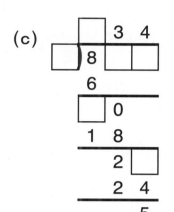

```
        □ 3 4
    □ ) 8 □ □
        6
      □ 0
      1 8
        2 □
        2 4
          5
```

(d)
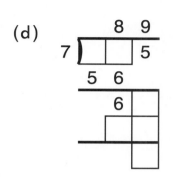

```
          8 9
    7 ) □ □ 5
        5 6
          6 □
          □ □
```

4 2 identical bikes cost $396.
How much do 7 such bikes cost?

5 Emiliano wants to pack 320 candles equally into 6 identical boxes with the fewest number of candles left over.
How many candles will be left over?

6 Brianna has twice as much money as Amanda.

Catalina has twice as much money as Brianna.

Altogether, they have $595.

How much money does Catalina have?

Challenge

7 The product of the two numbers in the overlapping squares is the third number in each large square.

Fill in the missing numbers.

(a)

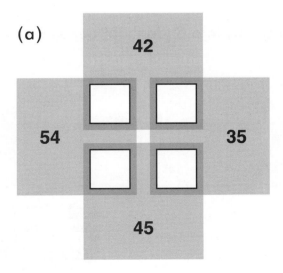

(b)

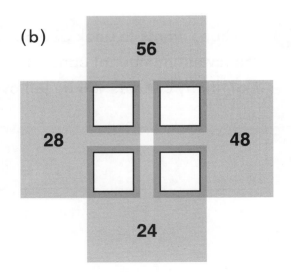

8-5 Practice A

Basics

1 $6 \times 8 = 8 \times 6 =$ ⬜

2 $9 \times 8 = 10 \times 8 -$ ⬜ $=$ ⬜

3 $7 \times 8 =$ ⬜ $+ 16 =$ ⬜

⬜ 2

4 (a) $8 \xrightarrow{\times 2}$ ⬜ $\xrightarrow{\times 2}$ ⬜ $\xrightarrow{\times 2}$ ⬜

$8 \times 8 =$ ⬜

(b) $4 \xrightarrow{\times 2}$ ⬜ $\xrightarrow{\times 2}$ ⬜ $\xrightarrow{\times 2}$ ⬜

$4 \times 8 =$ ⬜

5 (a) ⬜ $\times 8 = 72$ | $72 \div 8 =$ ⬜

(b) ⬜ $\times 8 = 24$ | $24 \div 8 =$ ⬜

6 $64 \div 8 =$ ⬜ | $64 +$ ⬜ $= 70$

$70 \div 8$ is ⬜ with a remainder of ⬜ .

Practice

 7 Multiply or divide.

10 × 8 = ☐	2 × 8 = ☐	72 ÷ 8 = ☐
16 ÷ 8 = ☐	64 ÷ 8 = ☐	40 ÷ 8 = ☐
3 × 8 = ☐	32 ÷ 8 = ☐	80 ÷ 8 = ☐
56 ÷ 8 = ☐	7 × 8 = ☐	0 × 8 = ☐
4 × 8 = ☐	8 ÷ 8 = ☐	6 × 8 = ☐
5 × 8 = ☐	24 ÷ 8 = ☐	8 × 8 = ☐
20 × 8 = ☐	48 ÷ 8 = ☐	9 × 8 = ☐

How many teeth do adult elephants have?

Color the spaces that contain the answers to find out.

38	3	80	18	0	17	65
24	17	6	44	4	95	29
11	14	10	34	16	12	35
28	32	150	84	72	5	64
8	51	22	33	160	99	7
40	1	2	580	9	56	48
19	38	13	95	27	50	85

8 (a) 65 ÷ 8 is [ ] with a remainder of [ ].

(b) 50 ÷ 8 is [ ] with a remainder of [ ].

(c) 60 ÷ 8 is [ ] with a remainder of [ ].

(d) 22 ÷ 8 is [ ] with a remainder of [ ].

9 A store owner has 75 toys to display.
He displays 8 toys on each shelf, and the remaining toys on one more shelf.
How many shelves does he use?

Challenge

10 Caden has 3 times as many toy dinosaurs as Leo.
Ella has 5 fewer toy dinosaurs than Caden.
Altogether, they have 44 toy dinosaurs.
How many toy dinosaurs does Ella have?

Basics

1 (a) $6 \times 9 = 60 - 6 = \boxed{}$

(b) $7 \times 9 = 70 - \boxed{} = \boxed{}$

(c) $8 \times 9 = \boxed{} - 8 = \boxed{}$

(d) $9 \times 9 = 90 - \boxed{} = \boxed{}$

2 (a) $\boxed{} \times 9 = 45$

(b) $6 \times 9 = 45 + 10 - \boxed{} = \boxed{}$

3 (a) If you add the value of the digits in the products for the multiplication table of 9, when multiplied by numbers up to 10, the sum is _____.

(b) Circle the numbers that cannot be products of 9 and a whole number.

37   **27**   **36**   **54**   **62**   **56**

4 (a) $\boxed{} \times 9 = 81$ │ $81 \div 9 = \boxed{}$

(b) $\boxed{} \times 9 = 63$ │ $63 \div 9 = \boxed{}$

5 $63 \div 9 = \boxed{}$ │ $63 + \boxed{} = 69$

$69 \div 9$ is $\boxed{}$ with a remainder of $\boxed{}$.

Practice

6 Find the missing numbers.

$18 \div 9 =$ ☐ **N**	$8 \times 9 =$ ☐ **V**	$3 \times 9 =$ ☐ **P**
$4 \times 9 =$ ☐ **H**	$90 \div 9 =$ ☐ **L**	$54 \div 9 =$ ☐ **T**
$9 \times 9 =$ ☐ **D**	$9 \div 9 =$ ☐ **Y**	$7 \times 9 =$ ☐ **Y**
$9 \times$ ☐ $= 72$ **E**	$9 \times 2 =$ ☐ **E**	$81 \div 9 =$ ☐ **I**
$27 \div$ ☐ $= 9$ **U**	☐ $\times 9 = 63$ **Y**	$9 \times$ ☐ $= 45$ **N**
☐ $\div 9 = 5$ **T**	☐ $\div 6 = 9$ **L**	☐ $= 36 \div 9$ **S**

When scientists tried to find a replacement for rubber, what happened?

Write the letters that match the answers above to find out.

6	36	18	1	12	9	2	72	18	5	6	8	81

56	4	9	10	54	63	82	27	3	45	6	7	90

7 (a) $100 \div 9$ is ☐ with a remainder of ☐.

(b) $55 \div 9$ is ☐ with a remainder of ☐.

(c) $89 \div 9$ is ☐ with a remainder of ☐.

(d) $35 \div 9$ is ☐ with a remainder of ☐.

8 A deck of playing cards has 52 cards.
If as many cards as possible are dealt out equally to 9 players, how many cards does each player get?

Challenge

9 Tomas has $82 more than Franco.
If Tomas gives $5 to Franco, he will have 9 times as much money as Franco.
How much money do they have altogether?

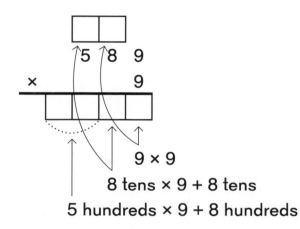

Basics

1 Fill in the missing numbers or digits.

$9 \times 9 =$ ☐

$80 \times 9 =$ ☐

$500 \times 9 =$ ☐

$589 \times 9 =$ ☐

$$\begin{array}{cccc} & 5 & 8 & 9 \\ \times & & & 9 \end{array}$$

← 9×9
← 80×9
← 500×9

$$\begin{array}{cccc} & 5 & 8 & 9 \\ \times & & & 9 \end{array}$$

9×9

8 tens $\times$ 9 + 8 tens

5 hundreds $\times$ 9 + 8 hundreds

2 $8 \times 8 =$ ☐

$50 \times 8 =$ ☐

$400 \times 8 =$ ☐

$458 \times 8 =$ ☐

$$\begin{array}{cccc} & 4 & 5 & 8 \\ \times & & & 8 \end{array}$$

Practice

3 Multiply.

890 × 8	9 × 64	8 × 71
	H	**A**
706 × 9	8 × 42	9 × 686
A	**N**	**O**
813 × 8	9 × 17	8 × 57
C	**R**	**N**

Riddle: When you need it, you throw it away.

When you are done with it, you bring it back.

What is it?

Write the letters that match the answers above to find out.

6,354	456	7,120	568	336	6,504	576	6,174	153

4 A deck of playing cards has 52 cards.
How many cards are in 9 decks?

5 9 boxes have 62 markers each.
A tenth box has 67 markers.
How many markers are there in all?

6 There are 9 teams with 5 students on each team.
Each student gets 8 cards for a game.
How many cards are needed for the game?

Basics

1 Divide 952 by 8.

8 ⟌ 9 5 2

1 5

7 2

0

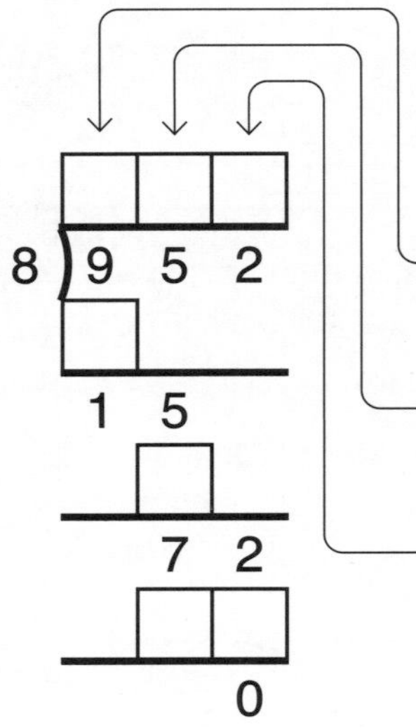

9 hundreds ÷ 8 is _____ hundred with 1 hundred left over.

15 tens ÷ 8 is _____ ten with 7 tens left over.

72 ones ÷ 8 is _____ ones with 0 ones left over.

952 ÷ 8 = ☐

Check: ☐ × 8 = 952

2 Divide 672 by 9.

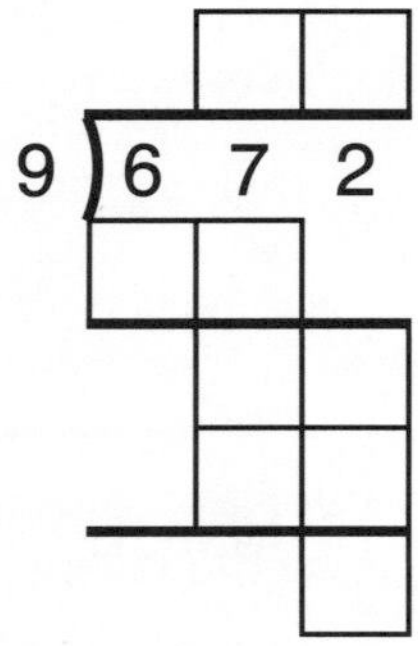

9 ⟌ 6 7 2

672 ÷ 9 is ☐ with a remainder of ☐ .

Check: ☐ × 9 + 6 = 672

Practice

3 Divide.

The sum of the remainders should equal the product of 9 and 2.

972 ÷ 8	473 ÷ 9	555 ÷ 9	683 ÷ 8

4 A rope is 458 m long.

It is cut into pieces that are each 8 m long.

How many pieces are there?

How long is the leftover piece?

5 A baker made 250 cookies and gave away 34 of them.

She put the rest equally into 9 tins.

How many cookies are in each tin?

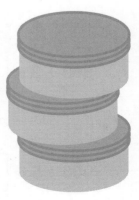

6 Find the missing numbers.
What pattern do you notice?

100 ÷ 9 is [ ] with a remainder of [ ].

200 ÷ 9 is [ ] with a remainder of [ ].

300 ÷ 9 is [ ] with a remainder of [ ].

400 ÷ 9 is [ ] with a remainder of [ ].

500 ÷ 9 is [ ] with a remainder of [ ].

600 ÷ 9 is [ ] with a remainder of [ ].

700 ÷ 9 is [ ] with a remainder of [ ].

800 ÷ 9 is [ ] with a remainder of [ ].

900 ÷ 9 is [ ] with a remainder of [ ].

Exercise 10

Check

1 The numbers 1 through 10 appear only once in each shaded row and column. Compete the multiplication table.

×										
					9					
				36						
								1		
							25			
		4								
	49									
										16
								81		
						100				
			64							

2 Multiply or divide.

9 × 847 **L**	680 ÷ 8 **E**	216 ÷ 9 **R**
792 ÷ 8 **S**	61 × 9 **K**	8 × 792 **A**
8 × 38 **A**	495 ÷ 9 **W**	974 × 8 **H**

What is the largest type of shark?
Write the letters that match the answers above to find out.

55	7,792	304	7,623	85	76	99	7,792	6,336	24	549

3 Write the missing digits.

(a)
```
      □ □ 7
    ×     8
  ─────────
    6 , 0 □ 6
```

(b)
```
      1 □ 5
    ×     9
  ─────────
    □ □ 2 5
```

(c)
```
          7 □
      ┌────────
    □ )6 □  9
      ─────────
       5  6
      ─────────
          4 □
      ─────────
          4  8
      ─────────
             □
```

(d)
```
          □  6
      ┌────────
    9 )□ □  7
      ─────────
       7  2
      ─────────
          5  □
      ─────────
          5  4
      ─────────
             □
```

4 An orchard has 8 times as many apple trees as pear trees.
There are 981 trees in all.
How many more apple trees than pear trees does it have?

5 A box holds 8 large bowls and 6 small bowls.
Wanda needs 250 large bowls for a restaurant.
How many of these boxes does she need to buy?
How many small bowls will she also have?

6 (a) $78 = \boxed{} \times 8 + 6$

(b) $47 = 7 \times \boxed{} + 5$

7 9 children are playing a game with 126 counters.
Each child gets the same number of counters.
2 children left and the counters were shared out again.
How many more counters does each child get now than before?

Challenge

8 A park has 9 trees planted in a row,
with 7 bushes between each tree.
How many bushes are there?

9 There are 12 spiders and beetles.
Each spider has 8 legs, and each beetle has 6 legs.
There are 88 legs altogether.
How many spiders are there?

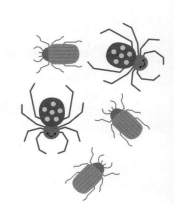

Chapter 9 Fractions — Part 1

Basics

1 A pentagon is divided into 5 equal parts.
3 parts are shaded.
$\frac{3}{5}$ of the pentagon is shaded.

(a) $\frac{3}{5}$ is ___3___ out of ___5___ equal parts.

(b) In $\frac{3}{5}$, the numerator is ___3___ and the denominator is ___5___.

(c) One part is $\boxed{\dfrac{3}{5}}$ of the whole.

(d) $\frac{3}{5}$ = ___3___ fifths

(e) 1 whole = ___5___ fifths

(f) $\frac{3}{5}$ and $\boxed{\dfrac{2}{5}}$ make 1 whole.

2

(a) On the bar, ___7___ out of ___10___ equal parts are shaded.

(b) $\boxed{\dfrac{7}{10}}$ of the bar is shaded.

(c) 1 whole = ___10___ tenths

(d) $\boxed{\dfrac{7}{10}}$ and $\boxed{\dfrac{3}{10}}$ make $\boxed{\dfrac{10}{10}}$, which is 1 whole.

Practice

3 What fraction of each shape is shaded?

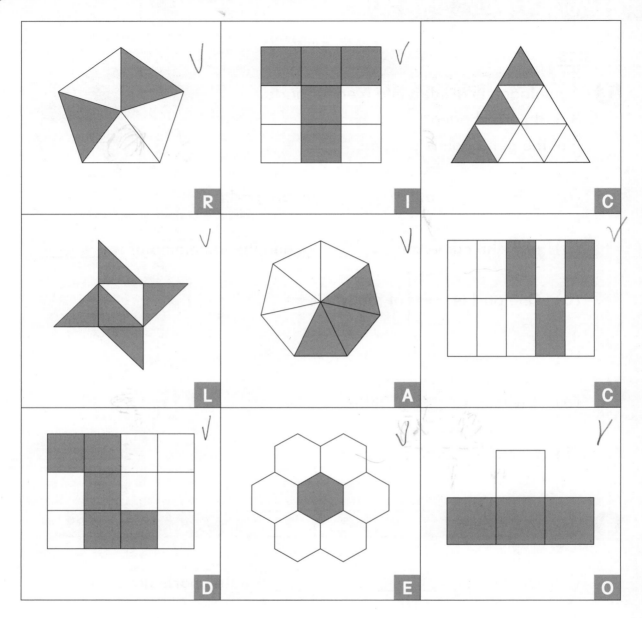

What animal cannot move its tongue?

Write the letters that match the answers above to find out.

	A		C	R	O	C	O	D	I	L	E	
$\frac{7}{12}$	$\frac{3}{7}$	$\frac{5}{8}$	$\frac{3}{9}$	$\frac{2}{5}$	$\frac{3}{4}$	$\frac{3}{10}$	$\frac{3}{4}$	$\frac{5}{12}$	$\frac{5}{9}$	$\frac{5}{6}$	$\frac{1}{7}$	$\frac{1}{6}$

4 Shade the given fraction for each shape.

(a) $\frac{6}{8}$

(b) $\frac{5}{10}$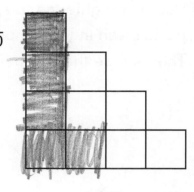

(c) $\frac{4}{12}$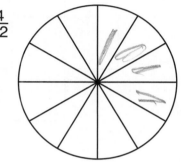

(d) $\frac{4}{9}$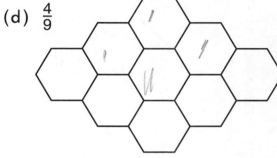

(e) $\frac{6}{9}$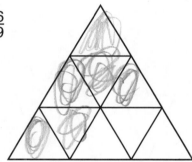

(f) $\frac{5}{8}$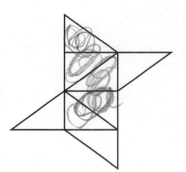

5 Join pairs of fractions that make 1.

| $\frac{5}{12}$ | $\frac{4}{7}$ | $\frac{7}{8}$ | $\frac{5}{7}$ | $\frac{6}{10}$ | $\frac{7}{12}$ | $\frac{5}{8}$ | $\frac{7}{15}$ |

| $\frac{3}{7}$ | $\frac{4}{10}$ | $\frac{7}{12}$ | $\frac{3}{8}$ | $\frac{1}{8}$ | $\frac{2}{7}$ | $\frac{8}{15}$ | $\frac{5}{12}$ |

Challenge

6 Use a straightedge to split each shape below into the number of equal parts given in the denominator of each fraction.

Then shade the number of parts given in the numerator of each fraction.

(a) $\frac{4}{5}$

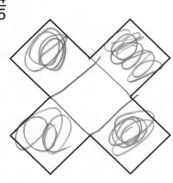

(b) $\frac{1}{2}$

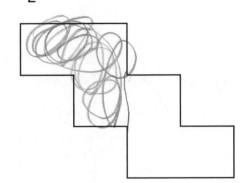

(c) $\frac{7}{12}$

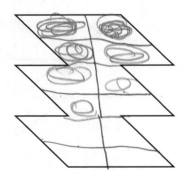

(d) $\frac{5}{7}$

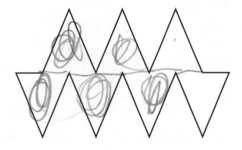

(e) $\frac{3}{4}$

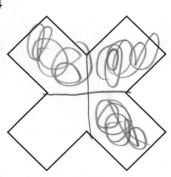

(f) $\frac{2}{3}$

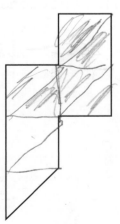

Basics

1 A one-meter tape has been divided into 5 equal parts.
3 parts are shaded.

1 m

(a) $\boxed{\dfrac{3}{}}$ of the tape is shaded.

(b) The shaded part is $\boxed{\dfrac{}{}}$ of a meter long.

(c) The rope is $\boxed{\dfrac{}{}}$ m long.

2 Answer the questions based on the number line.

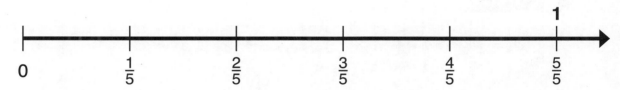

0     $\dfrac{1}{5}$ $\dfrac{2}{5}$ $\dfrac{3}{5}$ $\dfrac{4}{5}$ $\dfrac{5}{5}$

(a) The number line is divided into _____ equal parts between 0 and 1.

(b) There are _____ equal increments of $\frac{1}{5}$ between 0 and 1.

(c) $\frac{5}{5}$ is the same as _____.

(d) There are _____ equal increments of $\frac{1}{5}$ between 0 and $\frac{3}{5}$.

(e) There are 2 equal increments between $\boxed{\dfrac{}{5}}$ and 1.

(f) $\frac{5}{5}$ is $\frac{1}{5}$ and $\boxed{\dfrac{}{}}$.

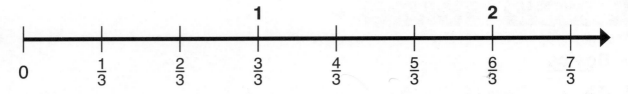

(a) The number line between 0 and 1 and also between 1 and 2 is divided into ___3___ equal parts.

(b) There are ___3___ thirds in $\frac{5}{3}$.

(c) There are ___6___ thirds in 2.

(d) $\frac{4}{3}$ is _____ thirds less than $\frac{7}{3}$.

Practice

 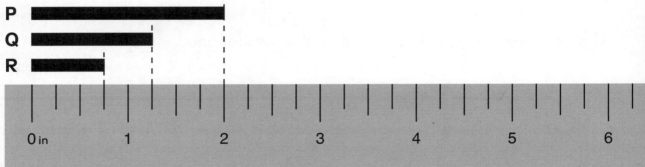

(a) This ruler shows increments of $\boxed{-}$ inch for each tick mark.

(b) Write the lengths of each line in fourths of an inch:

Line R: $\boxed{-}$ in   Line Q: $\boxed{-}$ in   Line P: $\boxed{-}$ in

(c) Line _____ is $\frac{2}{4}$ inches longer than line _____.

(d) Line R is $\boxed{\dfrac{-}{4}}$ inches shorter than line P.

5 Label the numbers marked with arrows on each number line.
Use those fractions to answer the questions below.

(a)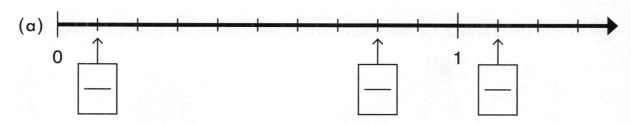

(b)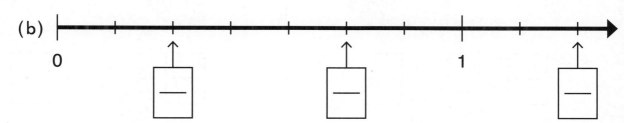

(c)

(d)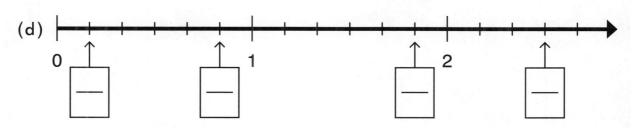

(e) List the fractions from problems (a) through (d) that are less than 1.

(f) List the fractions from problems (a) through (d) that are between 1 and 2.

Challenge

6 (a) _____ fifths make 5.   (b) _____ tenths make 100.

(c) $\boxed{\dfrac{\ \ }{3}}$ make 9.   (d) $\boxed{\dfrac{12}{\ \ }}$ make 3.

Basics

1 Write the fraction for the shaded part, then circle the greater fraction.

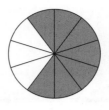  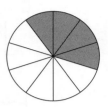

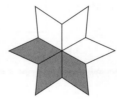 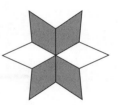

2 Write the fraction for the shaded part, then circle the lesser fraction.

3 Label the correct tick marks on each number line with the given fractions. Then circle the greatest fraction in each set.

(a) $\frac{7}{8}$ $\frac{3}{8}$ $\frac{5}{8}$

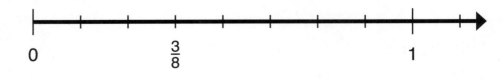

(b) $\frac{1}{3}$ $\frac{4}{3}$ $\frac{3}{3}$

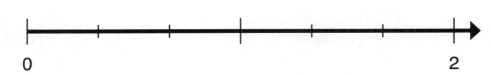

Practice

4 Write the fraction for the shaded part, then circle the greatest fraction.

5 Write the fraction for the shaded part, then circle the least fraction.

6 Label the correct tick marks on each number line with the given fractions. Then circle the least fraction in each set.

(a) $\frac{9}{10}$ $\frac{5}{10}$ $\frac{7}{10}$ $\frac{2}{10}$

0   1

(b) $\frac{7}{4}$ $\frac{3}{4}$ $\frac{8}{4}$ $\frac{5}{4}$

0   1

7

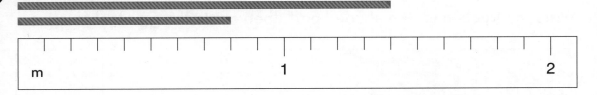

| | | | | | | | | | | | | | | | | | | | |
|m| | | | | | | |1| | | | | | | | |2| |

Write the lengths of the ropes in tenths of a meter in order, beginning with the shortest length.

8 Circle the greatest number.

(a) $\frac{2}{4}$ $\frac{3}{4}$ $\frac{5}{4}$ (b) $\frac{9}{10}$ $\frac{10}{10}$ $\frac{5}{10}$ $\frac{3}{10}$

9 Circle the least number.

(a) $\frac{9}{12}$ $\frac{7}{12}$ $\frac{2}{12}$ (b) $\frac{6}{7}$ $\frac{3}{7}$ $\frac{9}{7}$ $\frac{2}{7}$

10 Write the numbers in order, from least to greatest.

(a) $\frac{7}{6}, \frac{3}{6}, \frac{9}{6}, \frac{5}{6}$

(b) $\frac{9}{8}, \frac{3}{8}, 1, \frac{5}{8}, \frac{7}{8}$

11 Write the numbers in order, from greatest to least.

(a) $\frac{9}{12}, \frac{7}{12}, 1, \frac{5}{12}, 0$

(b) $1, \frac{3}{4}, \frac{5}{4}, 2, \frac{1}{4}$

Basics

1 Write the fraction for the shaded part, then circle the greater fraction.

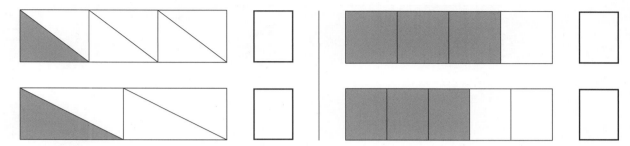

2 Write the fraction for the shaded part, then circle the lesser fraction.

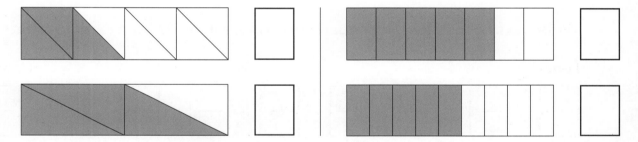

3 This number line shows 2 sets of tick marks.

(a) Label the tick marks marked with arrows.

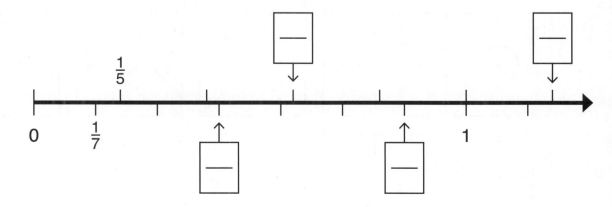

(b) Which is greater, $\frac{3}{5}$ or $\frac{3}{7}$?

(c) Which is less, $\frac{6}{5}$ or $\frac{6}{7}$?

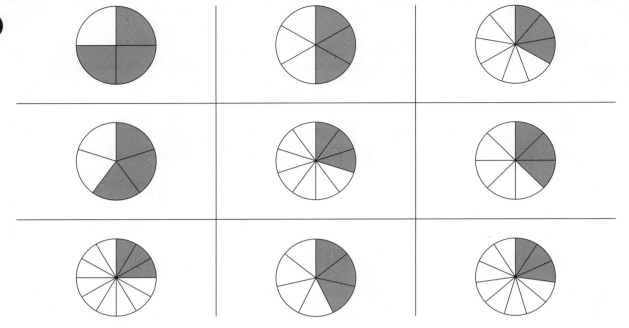

Write the fractions for the shaded parts above in order from least to greatest.

Least   **Greatest**

| Blue | Red | Yellow | White | Black | Blue | Black | Red | White |

Color the flags using the color code for each fraction above to see how to signal "I LOVE MATH" using maritime flags.

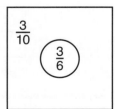 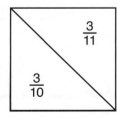 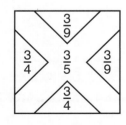 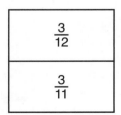

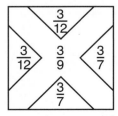 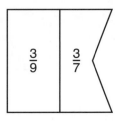 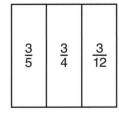 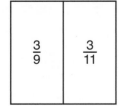

5 Color the bars to show the given fraction.
Then write > or < in each ◯.

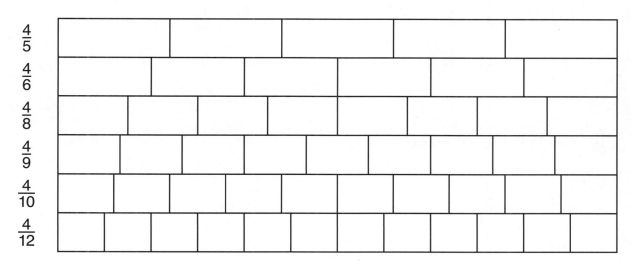

(a) $\frac{4}{8}$ ◯ $\frac{4}{5}$

(b) $\frac{4}{10}$ ◯ $\frac{4}{6}$

(c) $\frac{4}{9}$ ◯ $\frac{4}{12}$

6 Label the correct tick marks on the correct number lines with the fractions listed below.

Then write the fractions in order from least to greatest.

$\frac{8}{9}, \frac{8}{12}, \frac{8}{7}, \frac{8}{10}$

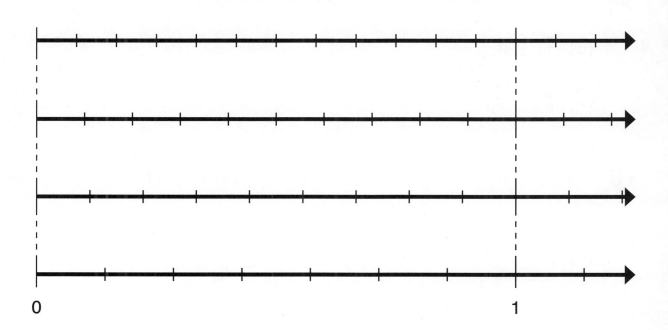

0   1

7 Circle the greatest number.

(a) $\frac{1}{5}$ $\frac{1}{8}$ $\frac{1}{4}$

(b) $\frac{7}{10}$ $\frac{7}{12}$ $\frac{7}{8}$ $\frac{7}{4}$

8 Circle the least number.

(a) $\frac{2}{8}$ $\frac{2}{2}$ $\frac{2}{5}$

(b) $\frac{9}{13}$ $\frac{9}{10}$ $\frac{9}{12}$ $\frac{9}{11}$

9 Write the numbers in order from least to greatest.

(a) $\frac{1}{5}, \frac{1}{2}, \frac{1}{3}$

(b) $\frac{5}{10}, \frac{5}{5}, \frac{5}{7}, \frac{5}{9}$

10 Write the numbers in order from greatest to least.

(a) $\frac{5}{12}, \frac{5}{5}, 0, \frac{5}{9}$

(b) $\frac{2}{10}, \frac{2}{5}, \frac{2}{15}, \frac{2}{9}, \frac{2}{7}$

11 Write a numerator or denominator that will make each of the following true.

(a) $\dfrac{5}{\boxed{}} < \dfrac{5}{7}$ (b) $\dfrac{3}{10} > \dfrac{\boxed{}}{12}$

Check

1 Multiply or divide.

68 × 8	365 × 4	589 × 7
97 ÷ 3	786 ÷ 5	634 ÷ 9

2 Which triangle correctly shows $\frac{3}{4}$ shaded?
Explain why the other triangles are not $\frac{3}{4}$ shaded.

A

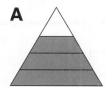

B

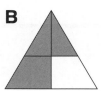

C

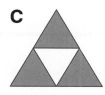

D

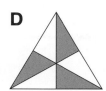

3 (a) $\boxed{\dfrac{}{}}$ of the triangle is not shaded.

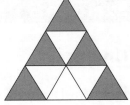

(b) $\dfrac{5}{9}$ and $\boxed{\dfrac{}{}}$ make 1.

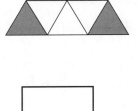

4 This figure was made using squares.
Shade $\dfrac{3}{5}$ of the figure.

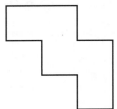

5 On which number line is the tick mark correctly labeled with the fraction?
Cross off the fractions that are incorrect and write the correct fraction.

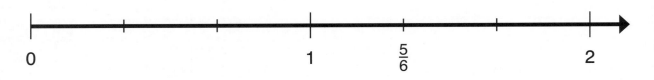

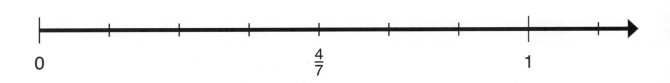

6 Label the correct tick marks on each number line with the given fractions. Then write the fractions in order from least to greatest.

(a) $\frac{7}{10}, \frac{7}{8}, \frac{3}{8}$

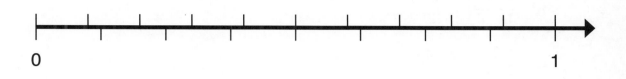

(b) $\frac{6}{5}, \frac{5}{3}, \frac{3}{5}, \frac{2}{3}$

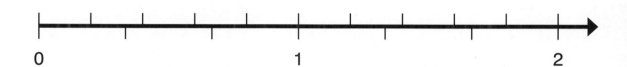

7 Write >, <, or = in each ◯.

(a) $\frac{3}{7}$ ◯ $\frac{6}{7}$

(b) $\frac{5}{4}$ ◯ $\frac{2}{4}$

(c) $\frac{4}{7}$ ◯ $\frac{4}{10}$

(d) 1 ◯ $\frac{2}{3}$

(e) $\frac{2}{2}$ ◯ $\frac{2}{11}$

(f) $\frac{6}{6}$ ◯ 6

8 Write a fraction that is greater than $\frac{4}{9}$ but less than $\frac{4}{7}$.

9 A pizza was cut into 8 equal pieces.
Gavin ate 2 pieces.
Franco ate $\frac{3}{8}$ of the pizza.
What fraction of the pizza is left?

Challenge

10 Write the numbers in order, from least to greatest.

(a) $\frac{7}{6}, \frac{3}{6}, \frac{3}{7}, \frac{2}{7}, \frac{5}{6}$

(b) $\frac{9}{12}, \frac{12}{8}, 1, \frac{5}{12}, \frac{9}{8}$

(c) $\frac{1}{5}, \frac{3}{3}, 3, \frac{8}{4}, \frac{1}{2}$

11 Match each fraction to the bar that is shaded by that amount.

$\frac{3}{6}$

$\frac{3}{5}$

$\frac{3}{4}$

$\frac{3}{7}$

12 How can you cut this cake into eighths with only 3 straight cuts?

Chapter 10 Fractions — Part 2

Basics

1 Write equivalent fractions for $\frac{1}{2}$.

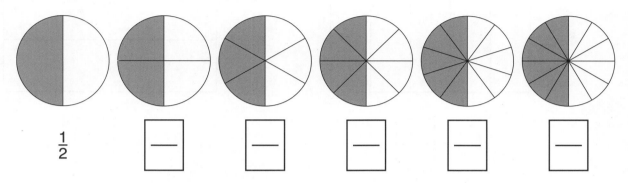

$\frac{1}{2}$ 　 $\boxed{}$ 　 $\boxed{}$ 　 $\boxed{}$ 　 $\boxed{}$ 　 $\boxed{}$

2 Write equivalent fractions for $\frac{3}{4}$.

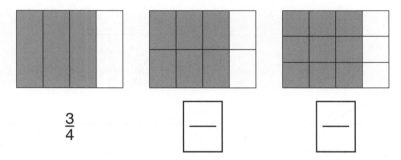

$\frac{3}{4}$ 　 $\boxed{}$ 　 $\boxed{}$

3 Use the number lines to find two equivalent fractions for $\frac{2}{3}$.

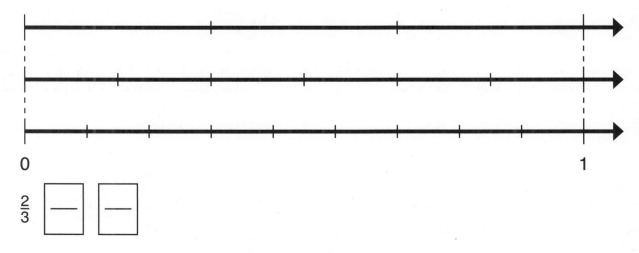

0 　　　　　　　　　　　　　　　　　　　　 1

$\frac{2}{3}$ $\boxed{}$ $\boxed{}$

Practice

4 Shade the given fraction on each shape.
Then write an equivalent fraction.

(a)

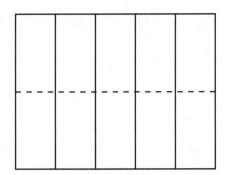

$$\frac{1}{5} = \boxed{\quad}$$

(b)

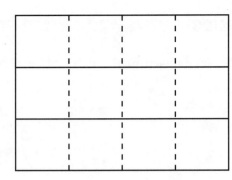

$$\frac{1}{3} = \boxed{\quad}$$

(c)

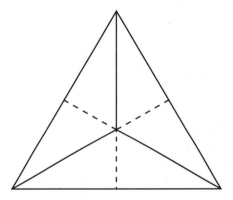

$$\frac{2}{3} = \boxed{\quad}$$

(d)

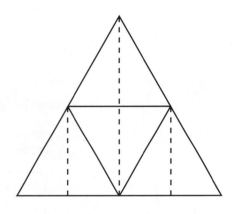

$$\frac{3}{4} = \boxed{\quad}$$

(e)

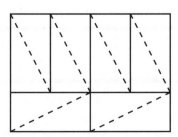

$$\frac{5}{6} = \boxed{\quad}$$

(f)

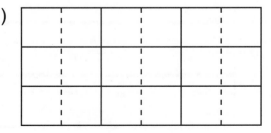

$$\frac{4}{9} = \boxed{\quad}$$

5 Label all pairs of equivalent fractions between 0 and 1 shown by the tick marks on each number line.
One has been done for you.

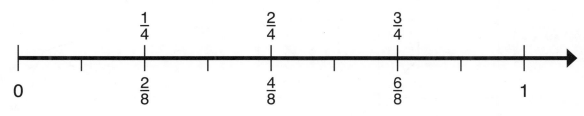

(a)

(b)

(c)

(d)

(e)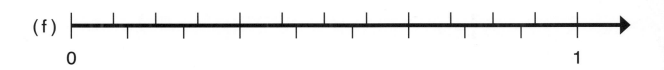

(f)

Challenge

6 Shade the given fraction on each figure.
Then write an equivalent fraction based on the figures.

(a)

(b)

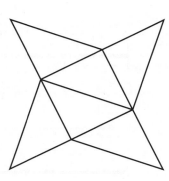

$$\frac{3}{8} = \boxed{}$$

$$\frac{1}{3} = \boxed{}$$

(c)

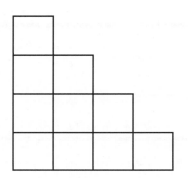

(d)

$$\frac{1}{2} = \boxed{}$$

$$\frac{2}{3} = \boxed{}$$

(e)

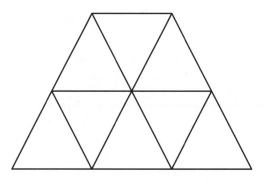

(f)

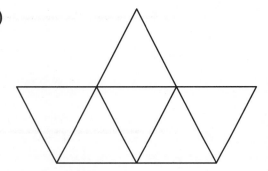

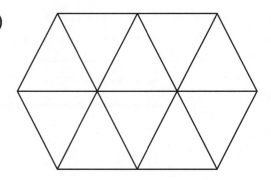

$$\frac{3}{4} = \boxed{}$$

$$\frac{3}{5} = \boxed{}$$

Basics

1 Write the missing numbers.

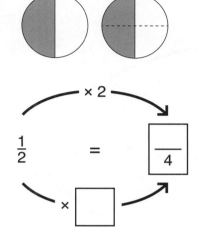

$$\frac{1}{2} = \frac{}{4}$$

× 2

×

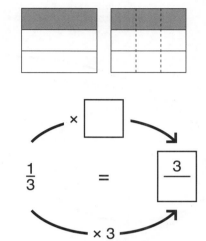

$$\frac{1}{3} = \frac{3}{}$$

×

× 3

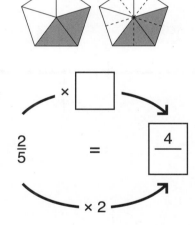

$$\frac{2}{5} = \frac{4}{}$$

×

× 2

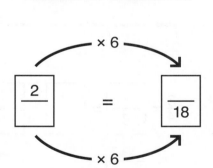

$$\frac{2}{} = \frac{}{18}$$

× 6

× 6

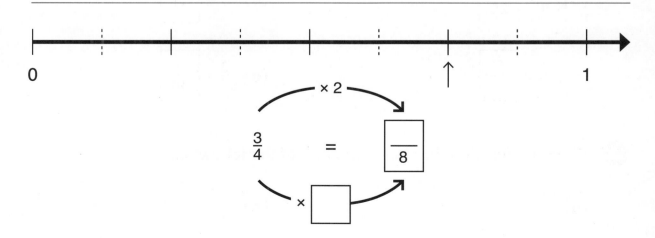

0   ↑   1

$$\frac{3}{4} = \frac{}{8}$$

× 2

×

Practice

2 Write the equivalent fractions.

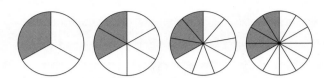

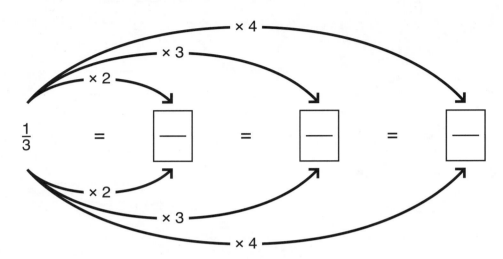

$$\frac{1}{3} = \boxed{\frac{}{}} = \boxed{\frac{}{}} = \boxed{\frac{}{}}$$

3 Write the missing numerators or denominators.

(a)

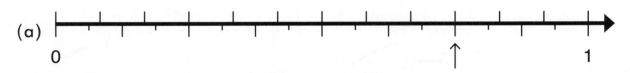

0   ↑   1

$$\frac{3}{4} = \boxed{\frac{}{8}} = \boxed{\frac{9}{}} = \boxed{\frac{}{16}}$$

(b) $\frac{3}{4} = \boxed{\dfrac{}{20}}$ (c) $\frac{2}{7} = \boxed{\dfrac{}{14}}$

(d) $\frac{2}{7} = \boxed{\dfrac{6}{}}$ (e) $\frac{3}{5} = \boxed{\dfrac{}{10}}$

4 Write 3 equivalent fractions for each of the following.

(a) $\frac{1}{5}$ (b) $\frac{4}{4}$

Challenge

5

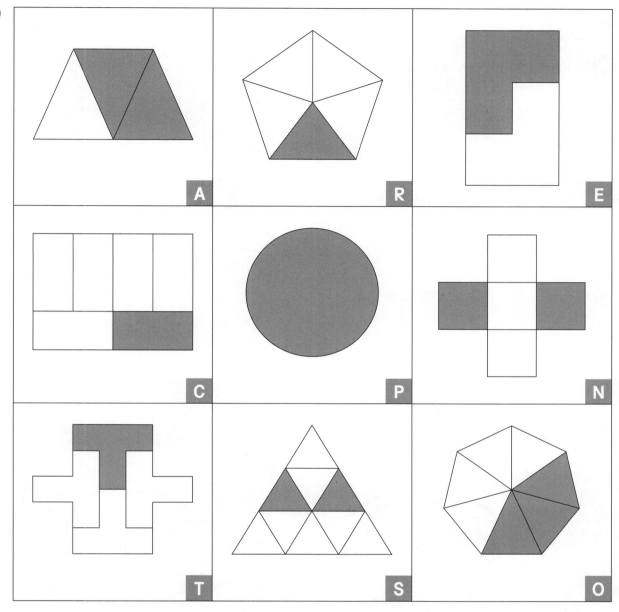

What dinosaur holds the record for the largest skull of all land animals ever?
Write the letters that match an equivalent fraction for the shaded part of
the figures above to find out.

	$\dfrac{5}{5}$	$\dfrac{4}{8}$	$\dfrac{4}{10}$	$\dfrac{2}{8}$	$\dfrac{6}{9}$	$\dfrac{2}{12}$	$\dfrac{3}{6}$	$\dfrac{4}{20}$	$\dfrac{8}{12}$	$\dfrac{3}{12}$	$\dfrac{6}{14}$	$\dfrac{11}{11}$	$\dfrac{4}{18}$

Basics

1 Write the missing numbers.

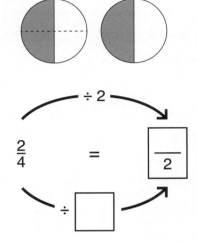

$$\frac{2}{4} = \frac{\boxed{}}{2}$$

÷ 2

÷ ☐

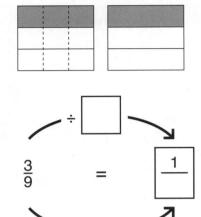

$$\frac{3}{9} = \frac{1}{\boxed{}}$$

÷ ☐

÷ 3

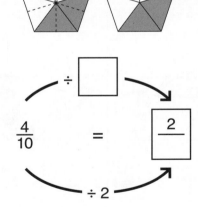

$$\frac{4}{10} = \frac{2}{\boxed{}}$$

÷ ☐

÷ 2

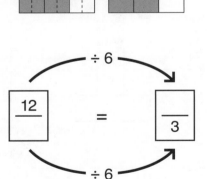

$$\frac{12}{\boxed{}} = \frac{\boxed{}}{3}$$

÷ 6

÷ 6

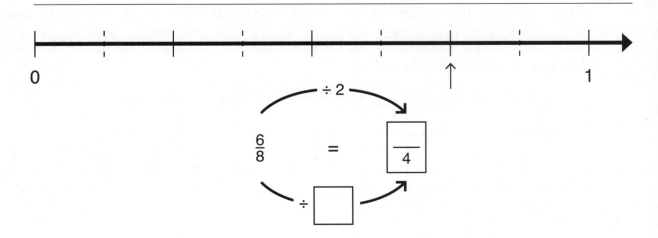

0   1

$$\frac{6}{8} = \frac{\boxed{}}{4}$$

÷ 2

÷ ☐

2 Simplify the fractions.
Write the missing numbers.

(a)

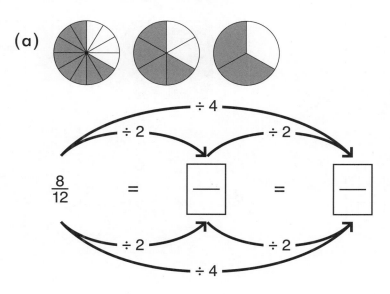

$$\frac{8}{12} = \boxed{} = \boxed{}$$

with arrows: ÷4 (top outer), ÷2 ÷2 (top inner); ÷2 ÷2 (bottom inner), ÷4 (bottom outer)

The simplest form of $\frac{8}{12}$ is $\boxed{}$.

(b)

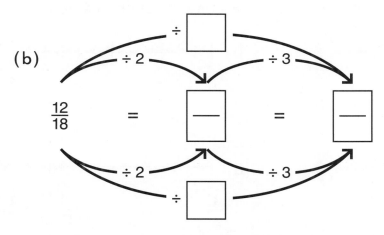

$$\frac{12}{18} = \boxed{} = \boxed{}$$

with arrows: ÷ □ (top outer), ÷2 ÷3 (top inner); ÷2 ÷3 (bottom inner), ÷ □ (bottom outer)

The simplest form of $\frac{12}{18}$ is $\boxed{}$.

(c)

0 ———————↑——— 1

$$\frac{12}{16} = \boxed{\dfrac{}{12}} = \boxed{\dfrac{6}{}} = \boxed{\dfrac{}{4}}$$

The simplest form of $\frac{12}{16}$ is $\boxed{}$.

Practice

3 Write the shaded fraction of each shape in simplest form.

(a)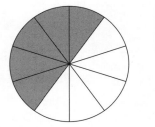

(b)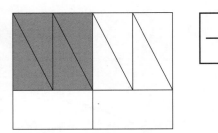

(c)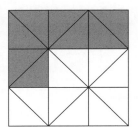

(d)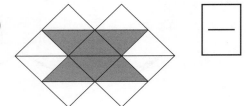

4 Label the tick marks indicated by the arrows as a fraction in simplest form.

(a)

(b)

(c)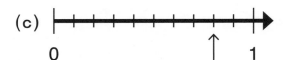

(d)

5 Write each fraction in simplest form.

(a) $\dfrac{6}{9} =$ [ ]

(b) $\dfrac{4}{10} =$ [ ]

(c) $\dfrac{8}{12} =$ [ ]

(d) $\dfrac{10}{20} =$ [ ]

6 How many muscles are in a cat's ear?

(a) Color the spaces with fractions in simplest form to find out.

$\frac{6}{9}$	$\frac{5}{6}$	$\frac{7}{10}$	$\frac{5}{9}$	$\frac{8}{10}$	$\frac{8}{15}$	$\frac{7}{8}$	$\frac{2}{11}$	$\frac{3}{9}$
$\frac{3}{15}$	$\frac{5}{10}$	$\frac{10}{12}$	$\frac{4}{5}$	$\frac{2}{4}$	$\frac{6}{8}$	$\frac{8}{16}$	$\frac{3}{17}$	$\frac{7}{14}$
$\frac{4}{20}$	$\frac{3}{10}$	$\frac{9}{20}$	$\frac{2}{3}$	$\frac{7}{7}$	$\frac{2}{5}$	$\frac{4}{9}$	$\frac{3}{4}$	$\frac{5}{20}$
$\frac{2}{10}$	$\frac{6}{12}$	$\frac{4}{8}$	$\frac{7}{12}$	$\frac{3}{6}$	$\frac{5}{13}$	$\frac{9}{15}$	$\frac{12}{16}$	$\frac{6}{15}$
$\frac{10}{16}$	$\frac{3}{14}$	$\frac{1}{2}$	$\frac{7}{15}$	$\frac{8}{12}$	$\frac{7}{10}$	$\frac{4}{7}$	$\frac{11}{18}$	$\frac{6}{8}$

(b) What fraction of the spaces are colored?

(c) Is that fraction in simplest form?

Challenge

7 Find the equivalent fractions.

(a)

$\frac{6}{9} \xrightarrow{\div 3} \boxed{} \quad \frac{6}{9} \xleftarrow{\div 3} $

$\frac{2}{3} \xrightarrow{\times 4} \boxed{} \quad \frac{2}{3} \xleftarrow{\times 4}$

$\frac{6}{9} = \boxed{\frac{}{12}}$

(b) $\frac{6}{8} = \boxed{\dfrac{}{12}}$

(c) $\frac{5}{10} = \boxed{\dfrac{}{16}}$

(d) $\frac{4}{6} = \boxed{\dfrac{}{9}}$

(e) $\frac{8}{10} = \boxed{\dfrac{}{15}}$

(f) $\frac{9}{15} = \boxed{\dfrac{}{10}}$

(g) $\frac{12}{16} = \boxed{\dfrac{}{20}}$

Basics

1 Color the bars to show each fraction.
Find the equivalent fraction.
Write > or < in each ◯.

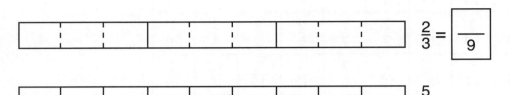

$$\frac{2}{3} = \boxed{\frac{}{9}}$$

$$\frac{5}{9}$$

$$\frac{2}{3} \bigcirc \frac{5}{9}$$

$$\frac{2}{5} = \boxed{\frac{4}{}}$$

$$\frac{4}{7}$$

$$\frac{2}{5} \bigcirc \frac{4}{7}$$

2 Label the given fractions on the number lines.
Find the equivalent fractions.
Write > or < in each ◯.

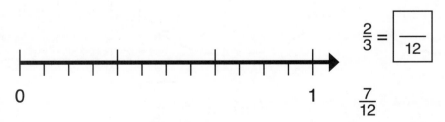

$$\frac{2}{3} = \boxed{\frac{}{12}}$$

$$\frac{7}{12}$$

$$\frac{2}{3} \bigcirc \frac{7}{12}$$

$$\frac{3}{5} = \boxed{\frac{9}{}}$$

$$\frac{9}{16}$$

$$\frac{3}{5} \bigcirc \frac{9}{16}$$

3 (a) Compare $\frac{2}{3}$ and $\frac{3}{5}$.

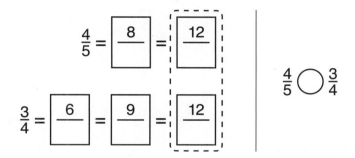

$$\frac{2}{3} = \frac{\boxed{}}{6} = \frac{\boxed{}}{9} = \frac{\boxed{}}{15}$$

$$\frac{3}{5} = \frac{\boxed{}}{10} = \frac{\boxed{}}{15}$$

$\frac{2}{3} \bigcirc \frac{3}{5}$

(b) Compare $\frac{4}{5}$ and $\frac{3}{4}$.

$$\frac{4}{5} = \frac{\boxed{8}}{} = \frac{\boxed{12}}{}$$

$$\frac{3}{4} = \frac{\boxed{6}}{} = \frac{\boxed{9}}{} = \frac{\boxed{12}}{}$$

$\frac{4}{5} \bigcirc \frac{3}{4}$

Practice

4 Write > or < in each $\bigcirc$.

(a) $\frac{7}{8} \bigcirc \frac{3}{4}$

(b) $\frac{3}{4} \bigcirc \frac{11}{12}$

(c) $\frac{2}{3} \bigcirc \frac{5}{9}$

(d) $\frac{3}{5} \bigcirc \frac{7}{15}$

(e) $\frac{4}{7} \bigcirc \frac{8}{9}$

(f) $\frac{3}{4} \bigcirc \frac{9}{11}$

(g) $\frac{1}{4} \bigcirc \frac{5}{12}$

(h) $\frac{3}{4} \bigcirc \frac{2}{3}$

5 Write the fractions in order from least to greatest.

(a) $\frac{5}{6}, \frac{2}{3}, \frac{1}{2}$

(b) $\frac{5}{12}, \frac{3}{4}, \frac{2}{3}$

(c) $\frac{5}{6}, \frac{2}{3}, \frac{7}{12}$

(d) $\frac{6}{11}, \frac{2}{3}, \frac{3}{10}$

(e) $\frac{8}{11}, \frac{1}{2}, \frac{4}{7}$

Challenge

6 Use the given numbers to fill in the missing numerators or denominators so that the fractions are in order from least to greatest.
Each fraction should be less than 1, and in simplest form.

(a) 1, 2, 5 $\boxed{\dfrac{}{3}} < \boxed{\dfrac{}{9}} < \boxed{\dfrac{}{3}}$

(b) 7, 8, 11 $\boxed{\dfrac{5}{}} < \boxed{\dfrac{5}{}} < \boxed{\dfrac{10}{}}$

(c) 3, 4, 7 $\boxed{\dfrac{1}{}} < \boxed{\dfrac{}{12}} < \boxed{\dfrac{3}{}}$

Basics

1 Color the bars to show each fraction and compare to $\frac{1}{2}$ or 1. Write >, <, or = in each $\bigcirc$.

(a)

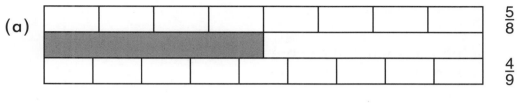

$\frac{5}{8}$

$\frac{4}{9}$

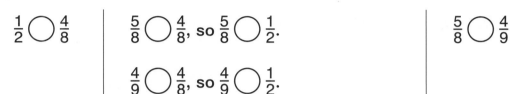

$\frac{1}{2}\bigcirc\frac{4}{8}$ | $\frac{5}{8}\bigcirc\frac{4}{8}$, so $\frac{5}{8}\bigcirc\frac{1}{2}$. | $\frac{5}{8}\bigcirc\frac{4}{9}$

$\frac{4}{9}\bigcirc\frac{4}{8}$, so $\frac{4}{9}\bigcirc\frac{1}{2}$.

(b)

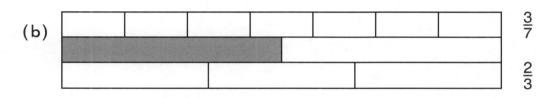

$\frac{3}{7}$

$\frac{2}{3}$

$\frac{1}{2}\bigcirc\frac{7}{14}$ | $\frac{3}{7}\bigcirc\frac{6}{14}$, and $\frac{6}{14}\bigcirc\frac{7}{14}$, so $\frac{3}{7}\bigcirc\frac{1}{2}$. | $\frac{3}{7}\bigcirc\frac{2}{3}$

$\frac{2}{4}\bigcirc\frac{2}{3}$ | $\frac{2}{3}\bigcirc\frac{2}{4}$, so $\frac{2}{3}\bigcirc\frac{1}{2}$.

(c)

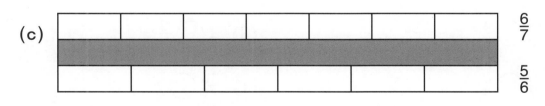

$\frac{6}{7}$

$\frac{5}{6}$

$\frac{6}{7}$ and $\frac{1}{7}$ make 1. | $\frac{1}{7}\bigcirc\frac{1}{6}$, so $\frac{6}{7}\bigcirc\frac{5}{6}$.

$\frac{5}{6}$ and $\frac{1}{6}$ make 1.

Practice

2 Circle all of the fractions that are less than $\frac{1}{2}$.

$$\frac{2}{6} \qquad \frac{4}{6} \qquad \frac{3}{8} \qquad \frac{5}{8} \qquad \frac{4}{10} \qquad \frac{6}{10} \qquad \frac{5}{12} \qquad \frac{7}{12} \qquad \frac{7}{16} \qquad \frac{9}{16}$$

3 Circle all of the fractions that are greater than $\frac{1}{2}$.

$$\frac{3}{5} \qquad \frac{3}{7} \qquad \frac{4}{7} \qquad \frac{4}{9} \qquad \frac{5}{9} \qquad \frac{5}{11} \qquad \frac{6}{11} \qquad \frac{6}{13} \qquad \frac{7}{13} \qquad \frac{7}{15}$$

4 Write > or < in each $\bigcirc$.

(a) $\frac{3}{8} \bigcirc \frac{5}{9}$

(b) $\frac{5}{8} \bigcirc \frac{7}{12}$

(c) $\frac{2}{3} \bigcirc \frac{5}{8}$

(d) $\frac{3}{7} \bigcirc \frac{8}{15}$

(e) $\frac{5}{7} \bigcirc \frac{3}{8}$

(f) $\frac{3}{4} \bigcirc \frac{5}{11}$

5 Write > or < in each $\bigcirc$.

(a) $\frac{7}{8} \bigcirc \frac{6}{7}$

(b) $\frac{4}{5} \bigcirc \frac{11}{12}$

(c) $\frac{2}{3} \bigcirc \frac{4}{5}$

(d) $\frac{6}{7} \bigcirc \frac{9}{10}$

(e) $\frac{6}{7} \bigcirc \frac{8}{9}$

(f) $\frac{10}{11} \bigcirc \frac{8}{9}$

6 Write the fractions in order from least to greatest.

(a) $\frac{5}{6}, \frac{6}{7}, \frac{2}{5}$

(b) $\frac{7}{12}, \frac{3}{4}, \frac{3}{7}$

(c) $\frac{5}{9}, \frac{2}{5}, \frac{13}{12}$

(d) $\frac{9}{11}, \frac{3}{7}, \frac{3}{10}, \frac{8}{8}$

(e) $\frac{8}{9}, \frac{1}{2}, \frac{3}{7}, \frac{7}{8}$

Challenge

7 Use the given numbers to fill in the missing numerators or denominators so that the fractions are in order from least to greatest.
Each fraction should be less than 1.

(a) 7, 8, 9

$$\frac{4}{} < \frac{7}{} < \frac{8}{}$$

(b) 2, 3, 4

$$\frac{}{7} < \frac{}{9} < \frac{}{4}$$

(c) 6, 7, 8

$$\frac{3}{} < \frac{}{9} < \frac{}{7}$$

Check

1 Which figure does not correctly show $\frac{3}{4}$ shaded?
Explain why the other figures are $\frac{3}{4}$ shaded.

A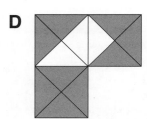

B

C

D

2 Label all pairs of equivalent fractions between 0 and 1 shown by the tick marks on each number line.

(a)

0   1

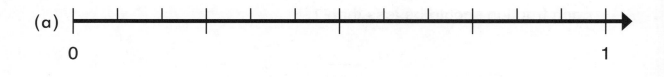

(b)

0   1

(c)

0   1

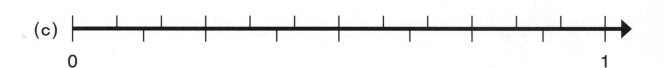

3 Circle all fractions greater than 1.

$$\frac{7}{9} \qquad \frac{6}{4} \qquad \frac{3}{3} \qquad \frac{8}{5} \qquad \frac{10}{9} \qquad \frac{7}{10} \qquad \frac{20}{16} \qquad \frac{15}{20}$$

4 Write 4 equivalent fractions for each of the following.

(a) $\frac{2}{3}$

(b) $\frac{3}{5}$

5 Fill in the numerators or denominators to write equivalent fractions for $\frac{1}{2}$.

(a) $\dfrac{6}{}$

(b) $\dfrac{}{6}$

(c) $\dfrac{5}{}$

(d) $\dfrac{}{16}$

6 Write each fraction in its simplest form.

(a) $\frac{6}{8} = \dfrac{}{}$

(b) $\frac{8}{10} = \dfrac{}{}$

(c) $\frac{9}{12} = \dfrac{}{}$

(d) $\frac{10}{16} = \dfrac{}{}$

7 $\frac{3}{10}$ of a bag of flour was used to bake bread and $\frac{1}{4}$ of the bag was used to make scones.

In which baked good was more flour used?

8 Label the tick marks on the number lines with the given fractions.

(a) $\frac{7}{8}, \frac{2}{3}, \frac{6}{7}, \frac{2}{5}$

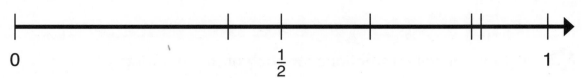

0        $\frac{1}{2}$ 1

(b) $\frac{3}{7}, \frac{8}{9}, \frac{3}{10}, \frac{5}{9}$

0        $\frac{1}{2}$ 1

9 Write the missing denominators.

(a) $\frac{1}{4} < \boxed{\dfrac{1}{}} < \frac{1}{2}$ (b) $\frac{2}{9} < \boxed{\dfrac{1}{}} < \frac{2}{7}$

Challenge

10 Write the missing denominator.

$\frac{2}{7} < \boxed{\dfrac{1}{}} < \frac{3}{8}$

11 The amount of water in a tank doubles each day.
If the tank becomes full on June 5, on what day was it half full?

Basics

1 Color each of the two fractions a different color on the figures.
Then add the fractions.

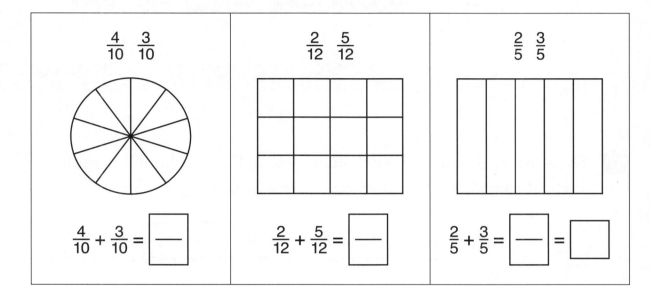

$$\frac{4}{10} \quad \frac{3}{10}$$

$$\frac{2}{12} \quad \frac{5}{12}$$

$$\frac{2}{5} \quad \frac{3}{5}$$

$$\frac{4}{10} + \frac{3}{10} = \boxed{}$$

$$\frac{2}{12} + \frac{5}{12} = \boxed{}$$

$$\frac{2}{5} + \frac{3}{5} = \boxed{} = \boxed{}$$

2 Show the addition on each number line, then add the fractions.
The first number line has been done for you.

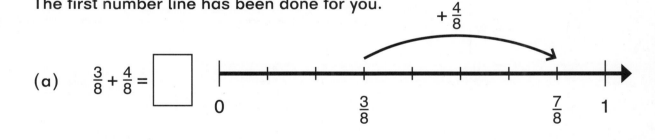

(a) $\frac{3}{8} + \frac{4}{8} = \boxed{}$

$+\frac{4}{8}$

0      $\frac{3}{8}$ $\frac{7}{8}$ 1

(b) $\frac{3}{7} + \frac{3}{7} = \boxed{}$

0   1

(c) $\frac{7}{16} + \frac{4}{16} = \boxed{}$

0   1

3 Use the bars to help you subtract.

(a) $\dfrac{8}{9} - \dfrac{4}{9} = \boxed{}$

(b) $\dfrac{8}{10} - \dfrac{5}{10} = \boxed{}$

(c) $1 - \dfrac{5}{6} = \boxed{}$

4 Show the subtraction on each number line, then subtract the fractions.
The first number line has been partly done for you.

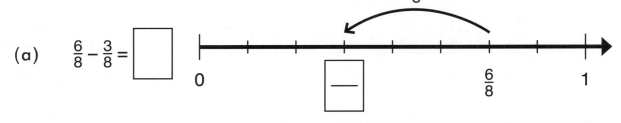

(a) $\dfrac{6}{8} - \dfrac{3}{8} = \boxed{}$

(b) $1 - \dfrac{4}{7} = \boxed{}$

(c) $\dfrac{11}{12} - \dfrac{6}{12} = \boxed{}$

(d) $\dfrac{3}{4} - \dfrac{3}{4} = \boxed{}$

Practice

5 Add or subtract.

(a) $\dfrac{1}{7} + \dfrac{3}{7} = \boxed{}$

(b) $\dfrac{3}{6} + \dfrac{3}{6} = \boxed{}$

(c) $\dfrac{8}{9} - \dfrac{7}{9} = \boxed{}$

(d) $\dfrac{10}{16} - \dfrac{5}{16} = \boxed{}$

(e) $1 - \dfrac{5}{12} = \boxed{}$

(f) $\dfrac{3}{9} + \dfrac{1}{9} + \dfrac{3}{9} = \boxed{}$

(g) $\dfrac{1}{4} + \dfrac{1}{4} + \dfrac{1}{4} + \dfrac{1}{4} = \boxed{}$

(h) $\dfrac{11}{15} - \dfrac{4}{15} - \dfrac{3}{15} = \boxed{}$

6 Ivy made some trail mix using $\dfrac{4}{8}$ cup of nuts, $\dfrac{2}{8}$ cup of raisins, and $\dfrac{1}{8}$ cup of chocolate chips.
How many cups of trail mix does she have in all?

7 Isaac has a board that is 1 meter long.
He sawed off a piece $\dfrac{4}{10}$ m long and another piece $\dfrac{3}{10}$ m long.
How long is the remaining piece?

Basics

1 Shade more units to add.
Cross off units to subtract.
Write each answer in simplest form.

(a)  $\dfrac{1}{6} + \dfrac{3}{6} = \boxed{\dfrac{}{}} = \boxed{\dfrac{}{}}$

(b) $\dfrac{2}{10} + \dfrac{3}{10} = \boxed{\dfrac{}{}} = \boxed{\dfrac{}{}}$

(c) $\dfrac{7}{9} - \dfrac{4}{9} = \boxed{\dfrac{}{}} = \boxed{\dfrac{}{}}$

(d) $\dfrac{11}{12} - \dfrac{3}{12} = \boxed{\dfrac{}{}} = \boxed{\dfrac{}{}}$

2 Show the addition or subtraction on each number line.
Write the answer in simplest form.

(a) $\dfrac{3}{8} + \dfrac{3}{8} = \boxed{\dfrac{}{}} = \boxed{\dfrac{}{}}$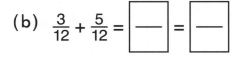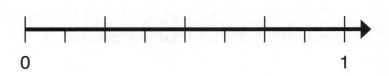

(b) $\dfrac{3}{12} + \dfrac{5}{12} = \boxed{\dfrac{}{}} = \boxed{\dfrac{}{}}$

(c) $\frac{7}{8} - \frac{3}{8} = \boxed{\dfrac{}{}} = \boxed{\dfrac{}{}}$

0   1

(d) $\frac{11}{16} - \frac{7}{16} = \boxed{\dfrac{}{}} = \boxed{\dfrac{}{}}$

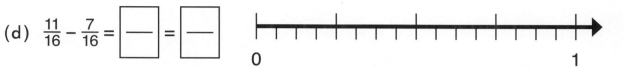

0   1

Practice

3 Write the missing numbers.

(a) $\boxed{\dfrac{1}{4}} + \boxed{\dfrac{}{4}} = \boxed{\dfrac{2}{}} = \boxed{\dfrac{1}{2}}$

(b) $\boxed{\dfrac{}{14}} + \boxed{\dfrac{7}{}} = \boxed{\dfrac{10}{}} = \boxed{\dfrac{5}{7}}$

(c) $\boxed{\dfrac{}{6}} - \boxed{\dfrac{2}{}} = \boxed{\dfrac{3}{6}} = \boxed{\dfrac{1}{}}$

(d) $\boxed{\dfrac{13}{20}} - \boxed{\dfrac{}{20}} = \boxed{\dfrac{4}{}} = \boxed{\dfrac{1}{5}}$

4 In a relay race, Aisha ran $\frac{5}{10}$ of a mile and Paula ran $\frac{3}{10}$ of a mile.

(a) How far did they run in all?

(b) Who ran farther, and how much farther?

5 Add or subtract.

$\dfrac{2}{10} + \dfrac{3}{10}$ **S**	$\dfrac{7}{8} - \dfrac{1}{8}$ **O**	$\dfrac{6}{7} - \dfrac{3}{7}$ **K**
$\dfrac{7}{8} - \dfrac{5}{8}$ **E**	$\dfrac{3}{9} + \dfrac{3}{9}$ **D**	$\dfrac{9}{14} + \dfrac{3}{14}$ **A**
$\dfrac{5}{18} + \dfrac{5}{18}$ **R**	$\dfrac{13}{15} - \dfrac{4}{15}$ **A**	$\dfrac{4}{9} - \dfrac{1}{9}$ **C**
$\dfrac{1}{5} + \dfrac{3}{5}$ **D**	$\dfrac{11}{20} - \dfrac{3}{20}$ **C**	$\dfrac{11}{16} + \dfrac{3}{16}$ **F**

Riddle: What is put on the table and cut in half, but never eaten?

Write the letters that match the values above to find out.

$\dfrac{3}{5}$	$\dfrac{3}{8}$	$\dfrac{4}{5}$	$\dfrac{1}{4}$	$\dfrac{2}{5}$	$\dfrac{3}{7}$	$\dfrac{1}{5}$	$\dfrac{3}{4}$	$\dfrac{7}{8}$	$\dfrac{5}{8}$	$\dfrac{1}{3}$	$\dfrac{6}{7}$	$\dfrac{5}{9}$	$\dfrac{2}{3}$	$\dfrac{1}{2}$

Check

 Find the values.

3,468 + 2,781	8,419 − 672	4,009 − 2,145
895 × 3	156 ÷ 5	665 × 7
96 ÷ 6	634 ÷ 9	888 × 2
230 + 987 + 158	1,569 + 3,871 + 78	45 + 2,409 + 9 + 348

2 How many more triangles must be shaded to show the given fractions? Complete the equations to show the sum of the shaded part and the part that still needs to be shaded.

(a)

Shade _____ more parts to have $\frac{2}{3}$ shaded.

$$\boxed{\frac{2}{9}} + \boxed{\frac{}{9}} = \frac{2}{3}$$

(b)

Shade _____ more parts to have $\frac{1}{2}$ shaded.

$$\boxed{\frac{}{6}} + \boxed{\frac{}{6}} = \frac{1}{2}$$

(c)

Shade _____ more parts to have $\frac{4}{5}$ shaded.

$$\boxed{\frac{3}{}} + \boxed{\frac{}{}} = \frac{4}{5}$$

(d)

Shade _____ more parts to have $\frac{6}{7}$ shaded.

$$\boxed{\frac{4}{}} + \boxed{\frac{}{}} = \frac{6}{7}$$

(e)

Shade _____ more parts to have $\frac{2}{3}$ shaded.

$$\boxed{\frac{5}{}} + \boxed{\frac{}{}} = \frac{2}{3}$$

(f)

Shade _____ more parts to have $\frac{3}{4}$ shaded.

$$\boxed{\frac{3}{}} + \boxed{\frac{}{}} = \frac{3}{4}$$

3 Complete the equations for the subtraction shown on each number line.

(a)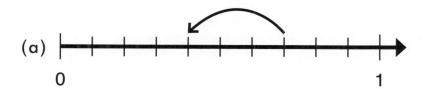

$$\boxed{\dfrac{7}{10}} - \boxed{\dfrac{3}{}} = \boxed{\dfrac{}{5}}$$

(b)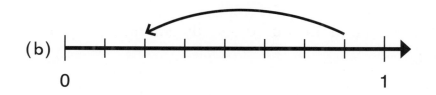

$$\boxed{\dfrac{}{}} - \boxed{\dfrac{}{}} = \boxed{\dfrac{}{4}}$$

(c)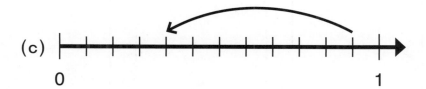

$$\boxed{\dfrac{}{}} - \boxed{\dfrac{}{}} = \boxed{\dfrac{}{3}}$$

4 Write >, <, or = in each $\bigcirc$.

(a) $\dfrac{1}{7} + \dfrac{6}{7} \bigcirc \dfrac{1}{9} + \dfrac{6}{9}$

(b) $\dfrac{4}{5} - \dfrac{4}{5} \bigcirc \dfrac{7}{8} - \dfrac{4}{8}$

(c) $\dfrac{1}{3} + \dfrac{1}{3} \bigcirc \dfrac{9}{12} - \dfrac{4}{12}$

(d) $\dfrac{1}{12} + \dfrac{3}{12} \bigcirc 1 - \dfrac{4}{6}$

(e) $\dfrac{3}{10} + \dfrac{2}{10} \bigcirc \dfrac{13}{16} - \dfrac{5}{16}$

(f) $\dfrac{5}{12} + \dfrac{5}{12} \bigcirc \dfrac{1}{4} + \dfrac{1}{4} + \dfrac{1}{4}$

(g) $\dfrac{1}{8} + \dfrac{3}{8} + \dfrac{1}{8} \bigcirc \dfrac{4}{5} - \dfrac{3}{5}$

(h) $1 - \dfrac{1}{3} \bigcirc \dfrac{2}{9} + \dfrac{2}{9}$

Challenge

5 Which is greater, $\frac{2}{5}$ or $\frac{7}{21}$?

6 Show how 4 chocolate bars can be shared equally among 5 people.

7 Use straight lines to divide this figure into fourths.
Each fourth should be the same shape and size.

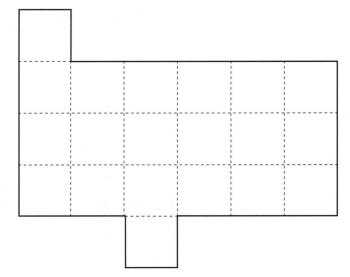

Chapter 11 Measurement

Basics

1 Fill in the blanks with m for meters or cm for centimeters.

(a) A baseball bat is about 1 _____ long.

(b) An 8-year-old child is about 125 _____ tall.

(c) A 3-story building is about 10 _____ tall.

(d) A hummingbird is about 10 _____ long.

2 (a) 1 m = 100 cm

2 m = ⬚ cm

6 m = ⬚ cm

10 m = ⬚ cm

(b) 100 cm = ⬚ m

200 cm = ⬚ m

500 cm = ⬚ m

900 cm = ⬚ m

3 A van is 8 m 15 cm long.
Find its length in centimeters.

8 m 15 cm = 800 cm + ⬚ cm

= ⬚ cm

The van is _____ cm long.

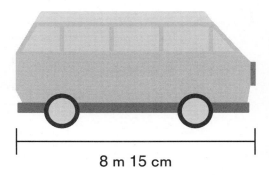

8 m 15 cm

4 A flag pole has a height of 740 cm.
Find its height in meters and centimeters.

740 cm = 700 cm + ⬚ cm

= 7 m ⬚ cm

The flag pole has a height of _____ m _____ cm.

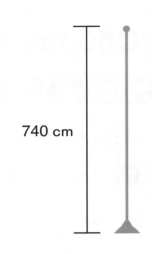

740 cm

5 A bed is 203 cm long.
Find its length in meters and centimeters.

203 cm = 200 cm + ⬚ cm

= ⬚ m ⬚ cm

The bed is _____ m _____ cm long.

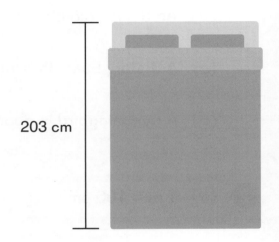

203 cm

6 Write the lengths in order, from shortest to longest.

3 m 2 cm, 203 cm, 320 cm, 2 m 30 cm

Practice

7 (a) 5 m 30 cm = ⬚ cm

(b) 6 m 8 cm = ⬚ cm

(c) 2 m 35 cm = ⬚ cm

(d) 7 m 27 cm = ⬚ cm

8 (a) 870 cm = ☐ m ☐ cm

(b) 525 cm = ☐ m ☐ cm

(c) 602 cm = ☐ m ☐ cm

9 (a) Circle the best estimate for the length of the computer monitor.

10 cm   **60 cm**

200 cm   **1 m 5 cm**

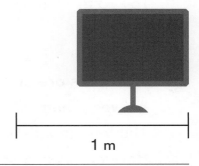

1 m

(b) Circle the best estimate for the height of the tree.

75 cm   **1 m 30 cm**

4 m   **2 m 90 cm**

200 cm   **105 cm**

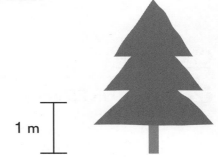

1 m

(c) Circle the best estimate for length of the car.

700 cm   **4 cm**

6 m   **45 cm**

4 m 50 cm   **6 m 20 cm**

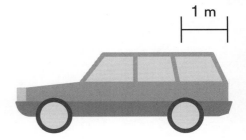

1 m

10 Write >, <, or = in each ◯.

(a) 4 m 25 cm ◯ 245 cm

(b) 101 cm ◯ 1 m 1 cm

(c) 570 cm ◯ 5 m 7 cm

(d) 8 m 18 cm ◯ 881 cm

11 Two sofas are put side by side with a gap of 5 cm between them.
The total length is 261 cm.
How long is one sofa in meters and centimeters?

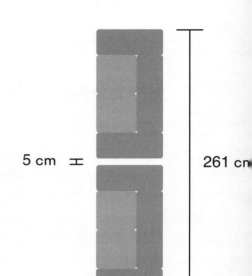

5 cm ⊣⊢   261 cm

Challenge

12 (a) 10 m 25 cm = [ ] cm

(b) 6,185 cm = [ ] m [ ] cm

13 A stack of 8 blocks has a height of 40 cm.
How many blocks are needed to make a stack that is 2 m high?

Basics

1 Draw straight lines to join each pair of lengths that add up to 1 m.
You will get 5 triangles.

43 cm
•

22 cm
•

41 cm •

• **34 cm**

• **6 cm**

• **59 cm**

94 cm •

•
78 cm

•
66 cm

•
57 cm

2 (a) 1 m − 41 cm = ☐ cm   2 m − 41 cm = 1 m ☐ cm

(b) 1 m − 22 cm = ☐ cm   5 m − 22 cm = 4 m ☐ cm

3 (a) 1 m − 6 cm = ☐ cm

(b) 6 m − 6 cm = ☐ m ☐ cm

(c) 1 m − 66 cm = ☐ cm

(d) 10 m − 66 cm = ☐ m ☐ cm

(e) 15 m − 78 cm = ☐ m ☐ cm

4 (a) 1 m − 43 cm = ☐ cm

(b) 7 m − 2 m = ☐ m

(c) 5 m − 43 cm = 4 m ☐ cm

(d) 7 m − 2 m 43 cm = 4 m ☐ cm

Practice

5 (a) 10 m − 8 m = ☐ m

(b) 2 m − 98 cm = ☐ m ☐ cm

(c) 10 m − 8 m 98 cm = ☐ m ☐ cm

6 (a) 1 m − 75 cm = ☐ cm

(b) 1 m − 82 cm = ☐ cm

(c) 2 m − 35 cm = ☐ m ☐ cm

(d) 2 m − 82 cm = ☐ m ☐ cm

(e) 8 m − 3 cm = ☐ m ☐ cm

7 (a) 35 cm + ☐ cm = 1 m   (b) 86 cm + ☐ cm = 1 m

(c) 8 cm + ☐ cm = 1 m   (d) ☐ cm + 28 cm = 1 m

8 (a) 6 m − 1 m 20 cm = [ ] m [ ] cm

(b) 9 m − 5 m 93 cm = [ ] m [ ] cm

(c) 4 m − 2 m 6 cm = [ ] m [ ] cm

(d) 9 m − 8 m 91 cm = [ ] cm

9 (a) 2 m 29 cm + [ ] cm = 3 m

(b) 5 m 62 cm + [ ] cm = 6 m

(c) 7 m [ ] cm + 18 cm = 8 m

(d) 1 m [ ] cm + 37 cm = 2 m

10 Rope A is 5 m 48 cm long and Rope B is 8 m long.
What is the difference in length between the two ropes?

11 A pole that is 7 m long is painted in two colors.
The first 3 m is painted green, and the last 65 cm is also painted green.
The middle portion is painted yellow.
What length of the pole is painted yellow?

Basics

1 Fill in the blanks with km for kilometers or m for meters.

(a) A bus is about 10 _____ long.

(b) Shanice went on a 6 _____ hike.

(c) The distance from Los Angeles to San Diego is about 180 _____.

(d) Mount Blackburn in Alaska is just under 5 _____ high.

(e) The Grand Canyon reaches a depth of about 1,850 _____.

2 (a) 1 km = 1,000 m   (b) 1,000 m = ☐ km

2 km = ☐ m   2,000 m = ☐ km

5 km = ☐ m   6,000 m = ☐ km

7 km = ☐ m   9,000 m = ☐ km

3 The height of Mount Everest is 8 km 848 m.
Write its height in meters.

8 km 848 m = 8,000 m + ☐ m

= ☐ m

Mount Everest has a height of _____ m.

4 The Hayes Volcano in Alaska has a height of 3,034 m above sea level. Find its height in kilometers and meters.

3,034 m = 3,000 m + ⬚ m

= ⬚ km ⬚ m

Hayes is _____ km _____ m high.

Practice

5 (a) 5 km 300 m = ⬚ m   (b) 2 km 205 m = ⬚ m

(c) 9 km 819 m = ⬚ m   (d) 6 km 80 m = ⬚ m

(e) 1 km 10 m = ⬚ m   (f) 7 km 7 m = ⬚ m

6 (a) 8,700 m = ⬚ km ⬚ m

(b) 9,147 m = ⬚ km ⬚ m

(c) 5,065 m = ⬚ km ⬚ m

(d) 6,002 m = ⬚ km ⬚ m

7 Write >, <, or = in each ◯.

(a) 2 km 520 m ◯ 2,450 m   (b) 8 km 18 m ◯ 818 m

(c) 5,100 m ◯ 5 km 1 m   (d) 5,070 m ◯ 5 km 70 m

8 This map shows the length of some trails near a campground.

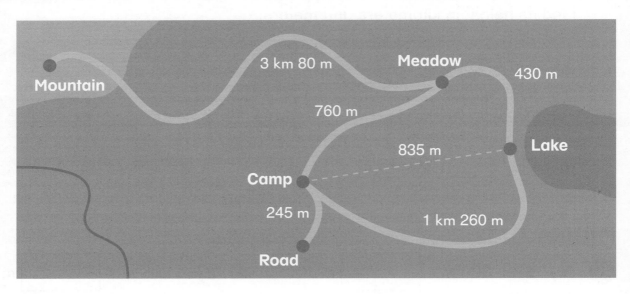

(a) What is the direct distance from the camp to the lake?

(b) What is the distance from the meadow to the mountain in meters?

(c) What is the total distance from the road to the meadow, passing through the camp, in kilometers and meters?

(d) What is the shortest distance from the camp to the lake along the trails in kilometers and meters?

Basics

1 (a) 1 km = 900 m + 90 m + [   ] m

(b) 700 m + [   ] m = 900 m

30 m + [   ] m = 90 m

4 m + [   ] m = 10 m

734 m + [   ] m = 1 km

2 Draw straight lines to join each pair of lengths that add up to 1 km.
You will get 2 pentagons and 1 triangle.

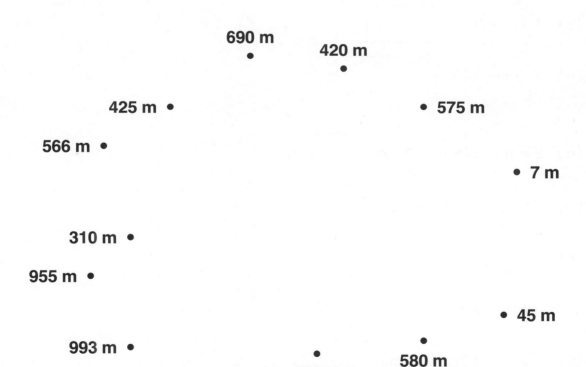

690 m

420 m

425 m

575 m

566 m

7 m

310 m

955 m

45 m

993 m

580 m

434 m

3 (a) 1 km − 690 m = ☐ m | 2 km − 690 m = 1 km ☐ m

(b) 1 km − 993 m = ☐ m | 5 km − 993 m = 4 km ☐ m

4 (a) 1 km − 420 m = ☐ m

(b) 3 km − 420 m = ☐ km ☐ m

(c) 9 km − 6 km = ☐ km

(d) 9 km − 6 km 420 m = ☐ km ☐ m

Practice

5 (a) 1 km − 650 m = ☐ m (b) 1 km − 652 m = ☐ m

(c) 1 km − 9 m = ☐ m (d) 1 km − 27 m = ☐ m

(e) 1 km − 349 m = ☐ m

6 (a) 2 km − 349 m = 1 km ☐ m

(b) 2 km − 1 km 700 m = ☐ m

(c) 2 km − 1 km 50 m = ☐ m

(d) 3 km − 1 km 5 m = ☐ km ☐ m

(e) 6 km − 3 km 470 m = ☐ km ☐ m

(f) 10 km − 3 km 47 m = ☐ km ☐ m

7 This map shows the length of some trails near a campground.

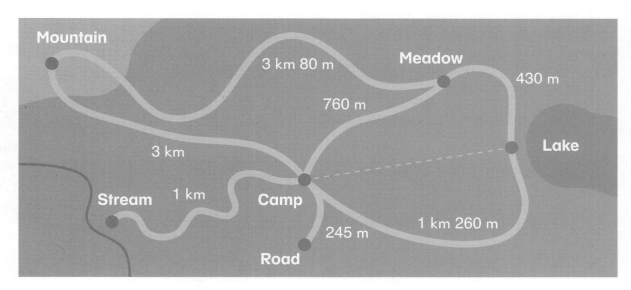

Find the differences in the length for the following trails, using the shortest possible routes.
Give your answers in compound units when possible.

(a) The camp to the stream and the meadow to the lake.

(b) The camp to the stream and the camp to the road.

(c) The camp to the mountain and the camp to the meadow.

(d) The camp to the mountain and the camp to the lake.

(e) The road to the meadow through the camp, and the camp to the mountain.

Basics

1 Fill in the blanks with L for liters or mL for milliliters.

(a) The capacity of a soda bottle is 1 _____ .

(b) A 1-cm cube could hold 1 _____ of water.

(c) A tablespoon holds about 15 _____ of water.

(d) A bottle of cooking oil holds about 1 _____ of oil.

(e) The capacity of a teacup is about 250 _____ .

2 (a) 1 L = 1,000 milliliters

(b) 1,000 mL = [____] L

3 L = [____] mL

6,000 mL = [____] L

3 How much water is in each beaker?

(a)

(b)

(c)

[____]

[____]

[____]

4

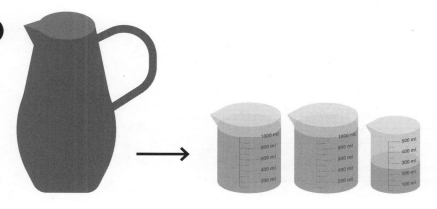

(a) Find the capacity of this container in milliliters.

1,000 mL + 1,000 mL + 250 mL = [＿＿＿] mL

(b) Find the capacity of this container in liters and milliliters.

1 L + 1 L + 250 mL = [＿＿＿] L [＿＿＿] mL

5 (a) 4,000 mL = [＿＿＿] L

(b) 4,200 mL = [＿＿＿] L [＿＿＿] mL

(c) 4,020 mL = [＿＿＿] L [＿＿＿] mL

(d) 4,002 mL = [＿＿＿] L [＿＿＿] mL

6 (a) 1 L − 230 mL = [＿＿＿] mL

(b) 2 L − 230 mL = [＿＿＿] L [＿＿＿] mL

(c) 8 L − 4 L 230 mL = [＿＿＿] L [＿＿＿] mL

Practice

 7 (a) 5 L 734 mL = ☐ mL

(b) 8 L 32 mL = ☐ mL

(c) 3,705 mL = ☐ L ☐ mL

(d) 6,043 mL = ☐ L ☐ mL

8 (a) 1 L – 345 mL = ☐ mL

(b) 7 L – 6 L 45 mL = ☐ mL

(c) 3 L – 3 mL = ☐ L ☐ mL

(d) 7 L – 2 L 380 mL = ☐ L ☐ mL

(e) 9 L – 4 L 75 mL = ☐ L ☐ mL

Challenge

9 How can we put 4 L in the bucket using only the 5-L and 3-L containers?

5 L

3 L

4 L

Basics

1 Fill in the blanks with kg for kilograms or g for grams.

(a) A paper clip weighs about 1 _____.

(b) A liter of water weighs 1 _____.

(c) An 8-year-old child weighs about 30 _____.

(d) A nickel weighs about 6 _____.

(e) An elephant can weigh 4,500 _____.

(f) A cat can weigh 4,500 _____.

2 (a) 1 kg = 1,000 grams

4 kg = [   ] g

(b) 1,000 g = [   ] kg

7,000 g = [   ] kg

3 (a) 6 kg 260 g = [   ] g

(b) 6 kg 26 g = [   ] g

4 (a) 4,900 g = [   ] kg [   ] g

(b) 4,009 g = [   ] kg [   ] g

5 (a) 1 kg – 430 g = [   ] g

(b) 8 kg – 430 g = 7 kg [   ] g

6 Write the weights of the following items.

(a)

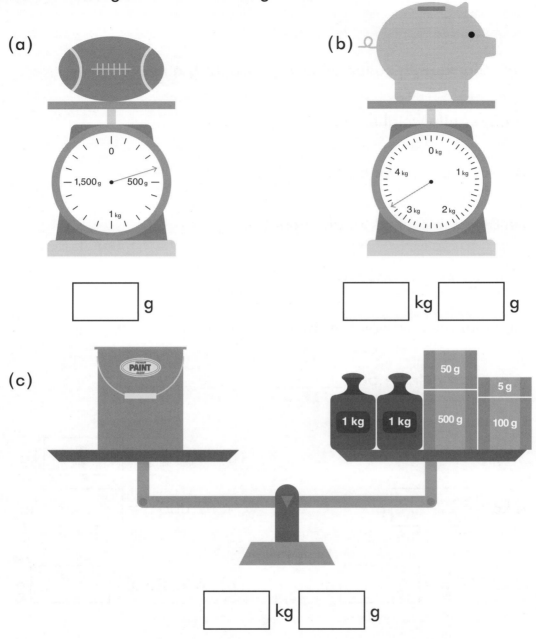

(b)

[   ] g

[   ] kg [   ] g

(c)

[   ] kg [   ] g

7 How much do these weights weigh altogether in kilograms and grams?

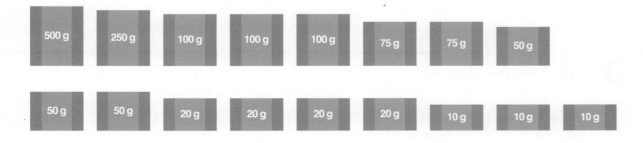

Practice

8 Write the following weights in order from lightest to heaviest.

8 kg 850 g

8,085 g

8,805 g

8,580 g

8 kg 58 g

8 kg 5 g

9 (a) 4 kg − 2 kg 900 g = [ ] kg [ ] g

(b) 8 kg − 1 kg 40 g = [ ] g

10 Write >, <, or = in each ◯.

(a) 9 kg 520 g ◯ 2,950 g

(b) 7,086 g ◯ 7 kg 68 g

(c) 4,050 g ◯ 5 kg − 250 g

(d) 4 kg − 789 g ◯ 4 kg − 250 g

(e) 4 kg − 3 kg 342 g ◯ 9 kg − 7 kg 342 g

(f) 8 kg − 3 kg 92 g ◯ 8 kg − 2 kg 92 g

(g) 7 kg − 3 kg 856 g ◯ 6 kg − 2 kg 568 g

Challenge

11 In each problem, the first and second scales balance.
How many triangles are needed to balance the third scale?

(a)

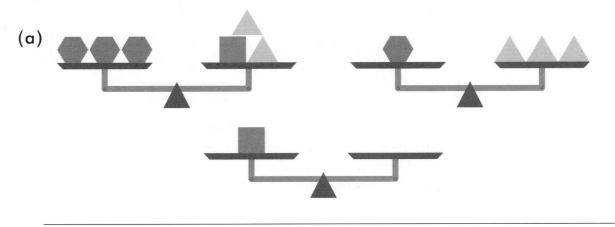

(b)

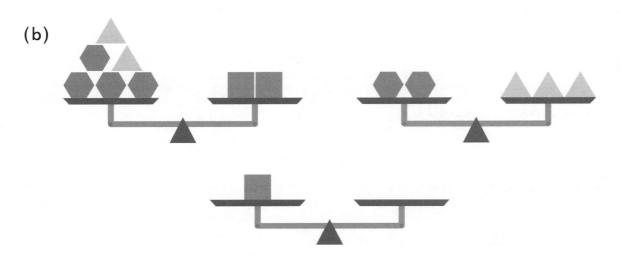

12 How much does B weigh in kilograms and grams?

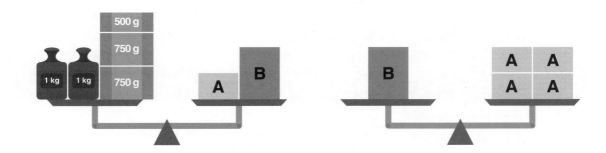

Basics

1 A tank can hold 4 L 300 mL of water.
It has 2 L 450 mL of water in it now.
How much more water is needed to fill it?

4 L 300 mL − 2 L 450 mL = [____] mL − [____] mL

= [____] mL

= [____] L [____] mL

_____ more water is needed to fill the tank.

2 1 bag of flour weighs 850 g.
How much do 6 such bags of flour weigh?

850 g × 6 = [____] g = [____] kg [____] g

6 bags of flour weigh _____.

3 9 bricks placed end to end are 2 m 7 cm long.
How long is 1 brick?

2 m 7 cm = [____] cm

[____] cm ÷ 9 = [____] cm

1 brick is _____ long.

4 The total weight of 3 tennis balls and 8 marbles is 380 g.

Each tennis ball weighs 60 g.

How much does 1 marble weigh?

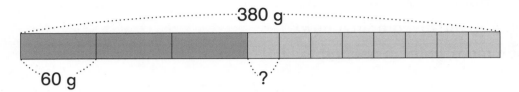

Practice

5 This map shows the length of some trails near a campground.

Write any answers greater than 1 km in compound units.

Assume the problem is asking for the shortest possible distances.

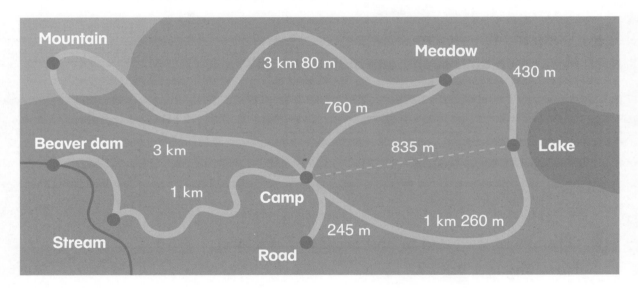

(a) How much longer is the distance along the trail from the camp to the
lake through the meadow than the direct distance to the lake?

(b) What is the total distance of the loop from the camp to the lake, then to the meadow, then to the mountain, and then back to the camp?

(c) Abigail went from the camp to the meadow and back twice. How far did she hike?

(d) Eli hiked from the camp past the lake to the meadow and then directly back to camp.
How much farther did Abigail hike than Eli?

(e) Maya hiked the trails from the camp to the beaver dam and back.
She hiked a total of 2 km 690 m.
How long is the trail from the stream to the beaver dam?

6 Santino cuts a 2-m rope into 8 equal pieces.
How long is each piece?

7 A container has a total capacity of 3 L 90 mL.
It was filled with 2 L 455 mL of water.
970 mL more water was poured into it, with the extra water overflowing.
How much water overflowed?

Challenge

8 There are two types of weights, A and B.
3 of weight A and 2 of weight B weigh 1 kg 300 g altogether.
1 of weight A and 2 of weight B weigh 840 g altogether.
How much do 1 of weight A and 1 of weight B weigh altogether?

Exercise 8

Check

1 Fill in the blanks with m, cm, km, g, kg, L, or mL.

(a) A hummingbird weighs about 4 _____.

(b) The lake is 100 _____ long.

(c) A dry bag for water sports has a capacity of 5 _____.

(d) A milk carton can hold 450 _____.

(e) 10 train cars hooked together are about 1 _____ long.

(f) The wingspan of an eagle is about 180 _____.

(g) An eagle weighs about 5 _____.

2 (a) 4 m 46 cm = [   ] cm

(b) 4 km 46 m = [   ] m

(c) 6 kg 207 g = [   ] g

(d) 2 L 2 mL = [   ] mL

(e) 4,689 g = [   ] kg [   ] g

(f) 8,020 m = [   ] km [   ] m

(g) 8,020 cm = [   ] m [   ] cm

3 4 L 250 mL of water weighs _____ grams.

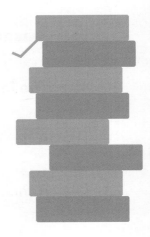

4 8 decks of cards weigh 768 g.
How much does one deck of cards weigh?

5 Javier Sotomayer has a high jump record of 2 m 45 cm.
Stefka Kostadinova has a high jump record of 209 cm.
Who has the greatest high jump record and by how much?

6 A cook had 6 bottles of olive oil, each of which contained 750 mL of olive oil.
After two weeks, she had 3 L 450 mL of olive oil left.
How much olive oil did she use in 2 weeks?
Give your answer in compound units.

7 This diagram shows the distance between some places in a town.

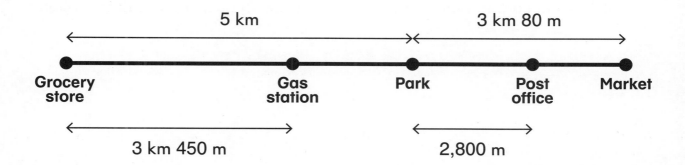

(a) Which two places are separated by 7,800 m?

(b) How much farther is it from the park to the market than from the park to the post office?

(c) Owen went from the park to the gas station and then to the post office. How far did he travel?

(d) What is the distance from the gas station to the market?

(e) Mila went from the gas station to 2 different places.
She traveled a total of 7 km 150 m.
Which 2 places did she go to and in what order?

Challenge

8 2 screwdrivers and 2 boxes of screws weigh 4 kg 632 g altogether.
If 1 box of screws weighs 2 kg 267 g, how much does 1 screwdriver weigh?

9 Two containers have 950 mL of water altogether.
Container B has 4 times as much water as container A.
How much water has to be poured from container B
into container A so they both have the same amount of water?

10 One cup is 16 cm tall.
Two stacked cups have a height of 20 cm.
How many cups are needed to make a stack that is 80 cm high?

Check

1 (a) How many tens make 9,400?

(b) How many tenths make 2?

(c) How many digits are there in the quotient of 581 ÷ 7?

(d) Estimate the sum of 4,586 + 768.

(e) In the sum of 3,985 and 1,567, what digit is in the tens place?

(f) For the difference between 42 and 6,003, what digit is in the hundreds place?

(g) What is the remainder for 689 ÷ 2?

2 Write the missing digits.

(a)
```
      □ 4 8 7
   + 2, 9 □ □
   ─────────
     9, □ 9 5
```

(b)
```
     6, 4 8 □
   − □ □ 0 8
   ─────────
     3, 5 □ 9
```

(c)
```
        6 2 □
   ×        7
   ─────────
     4, □ 0 3
```

(d)
```
          8 □
   8 ) □ □ 6
       6 4
       ───
         5 □
       □ □
       ───
          0
```

3 Write the numbers for all of the unlabeled tick marks shown on each number line.

(a)

0   500

(b)

0   1

(c)
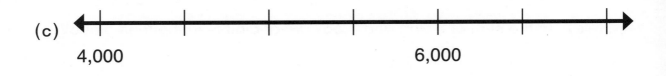
4,000   6,000

(d)
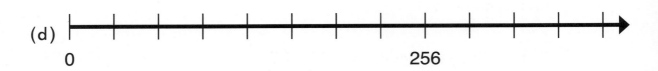
0   256

4 (a) $8 \times 8 = 16 + 16 + \boxed{}$

(b) $482 = \boxed{} \times 5 + 2$

(c) $620 \times 9 = 620 \times 10 - \boxed{}$

5 Write the fractions in simplest form.

(a) $\dfrac{10}{16} = \boxed{}$

(b) $\dfrac{6}{8} = \boxed{}$

(c) $\dfrac{25}{50} = \boxed{}$

6 The information below shows how far a paper airplane flew on 5 different trials.

Trial	1	2	3	4	5
Distance	1 m 80 cm	2 m 40 cm	90 cm	350 cm	260 cm

(a) Complete the bar graph with the information from the table.

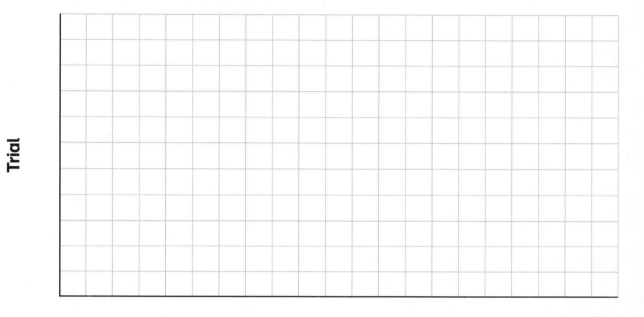

Trial

Distance (centimeters)

(b) Put the trials in order of least to greatest distance.

(c) What is the difference between the greatest and the least distance the plane flew?

7 The sum of three numbers is $\frac{7}{10}$.
The sum of the first and second numbers is $\frac{5}{10}$.
What is the third number?

8 n is a number. Find the value of n.

$$\frac{1}{n} = \frac{n}{16}$$

9 Is the product of $1 \times 2 \times 3 \times 4 \times .... \times 19 \times 20$ even or odd?

10 Sort the following fractions into 2 groups.
Explain why you grouped them that way.

$$\frac{1}{5} \qquad \frac{4}{5} \qquad \frac{3}{9} \qquad \frac{1}{3} \qquad \frac{2}{6} \qquad \frac{3}{5}$$

11 Fuyu used $\frac{2}{3}$ m of string to tie a package, $\frac{2}{5}$ m for an art project, and $\frac{3}{10}$ m to tie up a tomato plant.
On which of the above items did she use the most string?

12 (a) A men's shot put used in a track and field competition weighs 7 kg 260g. Write its weight in grams.

(b) Randy Barnes set a world record of 23 m 12 cm for the shot put throw. Write this distance in centimeters.

13 The total capacity of 3 containers, A, B, and C, is 9 L 700 mL altogether.
The total capacity of container A and B is 5 L 800 mL altogether.
The capacity container C is 800 mL more than the capacity of container A.
What is the capacity of each of the containers in liters and milliliters?

Challenge

14 There are 2 different trails to a lookout.
The total distance up and back for both trails is 900 m altogether.
Trail A is 20 m longer one way to the lookout than Trail B.
How long is Trail B one way?

15 At a booth at a fair, a 250 mL cup of lemonade was sold for $2.
If 60 liters of lemonade was sold in a day, how much money was received from selling lemonade that day?

16 There are 40 flags in a row.
The flags form a pattern of white, blue, yellow, white, blue, yellow, and so on.
The first flag on the left is white.
What color is the flag that is 6th from the right?

17 Find 2 fractions between $\frac{1}{3}$ and $\frac{1}{2}$.

Chapter 12 Geometry

Basics

1 (a) Trace the lines that are diameters.   (b) Trace the lines that are radii.

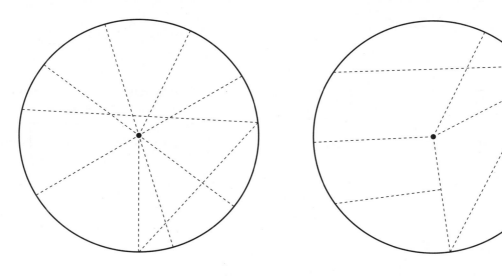

2 Measure the radius and diameter of each circle in centimeters.

(a)

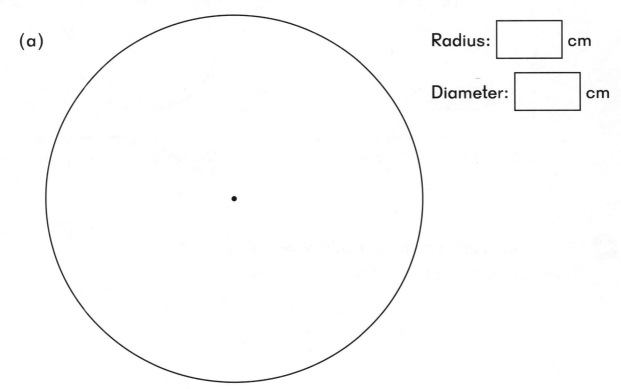

Radius: ☐ cm

Diameter: ☐ cm

(b)

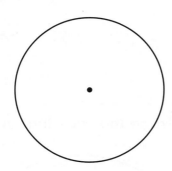

Radius: ☐ cm

Diameter: ☐ cm

(c)

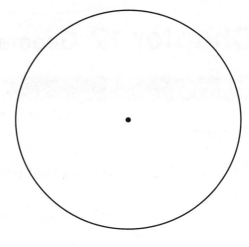

Radius: ☐ cm

Diameter: ☐ cm

Practice

3 How long is the diameter of each circle?

(a)

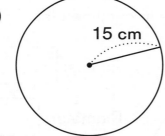

15 cm

Diameter: ☐ cm

(b)

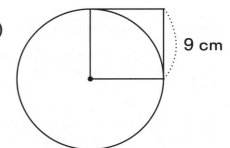

9 cm

Diameter: ☐ cm

4 A semi-circle has a straight side of 34 cm.
How long is the radius of the whole circle?

34 cm

5 How long is the radius of the each circle?

(a)

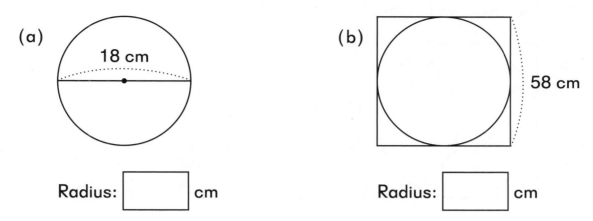

18 cm

Radius: [ ] cm

(b)

58 cm

Radius: [ ] cm

6 A quarter-circle has a straight side of 18 cm.
How long is the diameter of the circle?

7 All four circles in each problem are the same size.
Find the unknown lengths.

(a)

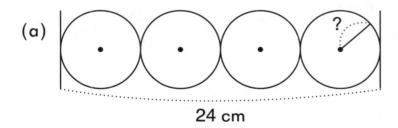

?

24 cm

(b)

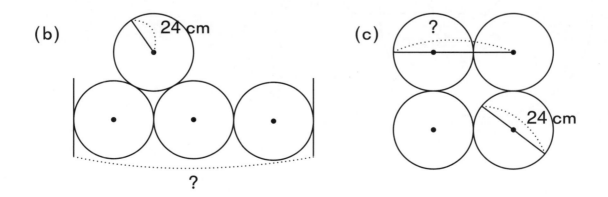

24 cm

?

(c)

?

24 cm

8 The diagrams show two circles and a rectangle.
Find the unknown lengths.

(a)

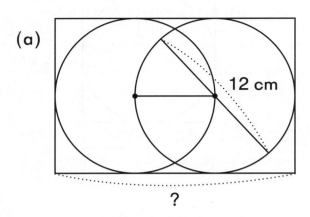

12 cm

?

(b)

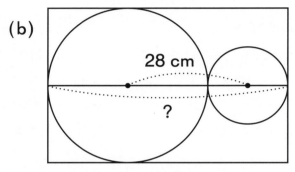

28 cm

?

Challenge

9 The diagram shows two circles.
Find the unknown length.

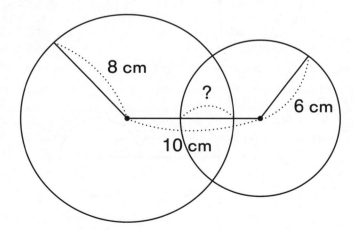

8 cm

?

6 cm

10 cm

Exercise 2

Basics

1 The diagram shows the angles formed between two strips of cardboard as one strip is opened away from the other.

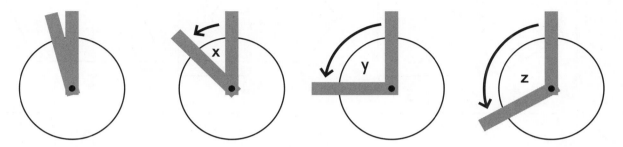

(a) Angle _____ is the largest.

(b) The vertex of each angle is at the _____ of the circle.

(c) The two lines are the _____ of the angle.

2 Check ✓ the pairs of lines that form angles.

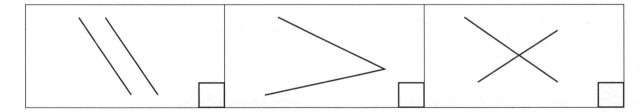

3 Check ✓ the figures that have at least one angle.

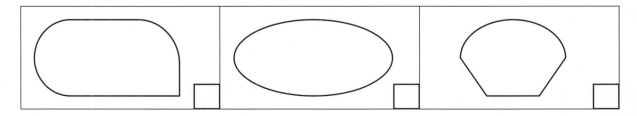

Practice

4 Mark inside angles on each figure and write the number of angles.
The first one has been done.

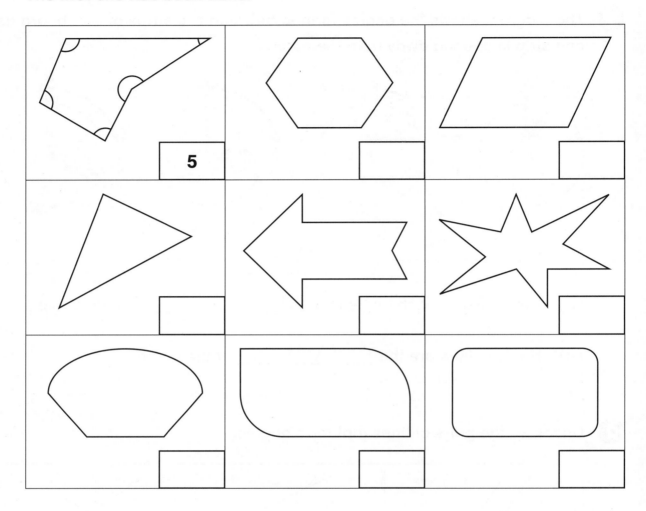

5 Use a ruler to draw a closed figure with straight sides.
How many inside angles does it have?

Basics

For any problem, you can use a set square or the corner of a rectangular card to compare the angles to a right angle.

1 All rectangles have _____ right angles.
How many right angles do each of the two triangles have?

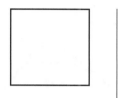 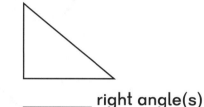 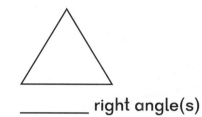

_____ right angle(s)   _____ right angle(s)

2 The diagram shows the angles formed between two strips of cardboard as one strip is opened away from the other.

 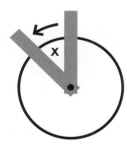 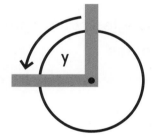 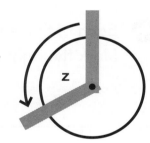

(a) Which of the labeled angles is smaller than a right angle?

(b) Which angle is a right angle?

(c) Which angle is larger than a right angle?

(d) Compare the angle on the right to the right angle above.
Is it larger, smaller, or the same size as a right angle?

Practice

3 Which of these angles are right angles?

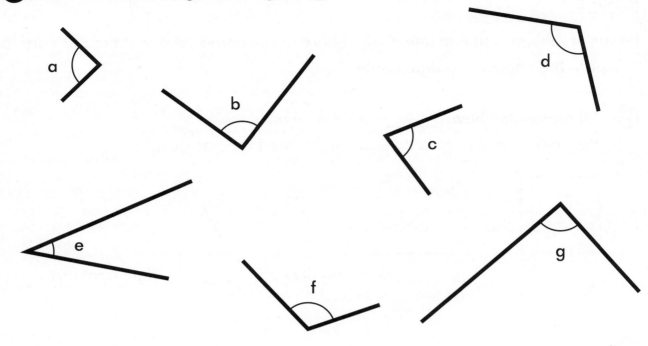

4 Each of these triangles has a right angle.
Mark the right angles.

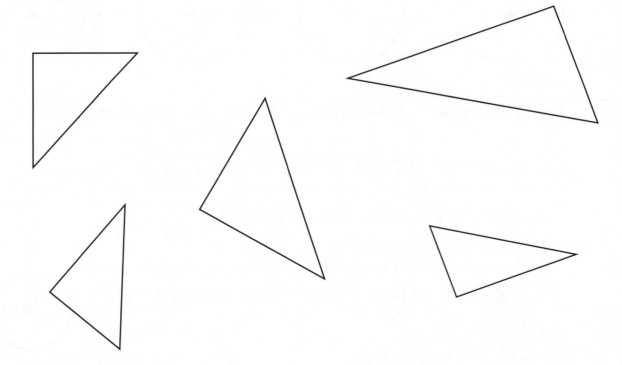

5

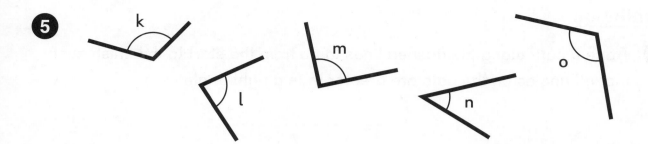

(a) Which angles are larger than a right angle?

(b) Which angles are smaller than a right angle?

(c) List the angles in order from smallest to largest.

6 Write how many right angles each figure has.
The right angles can be on the inside or on the outside of the figure.

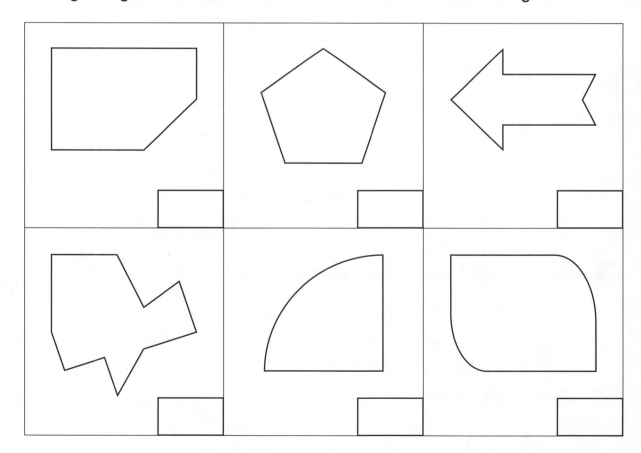

Challenge

7 Trace a path along the dashed lines to go from the start to the finish such that all angles on the path are greater than a right angle.

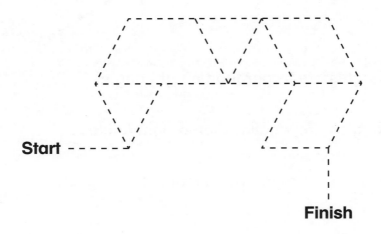

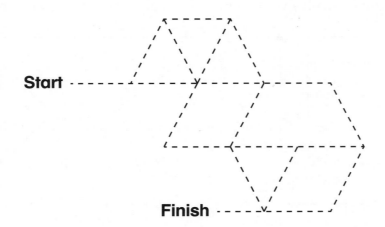

8 How many right angles does this figure have?

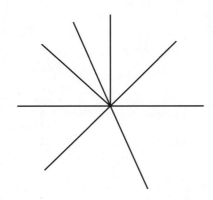

Basics

1 Measure the sides of the triangles in centimeters.

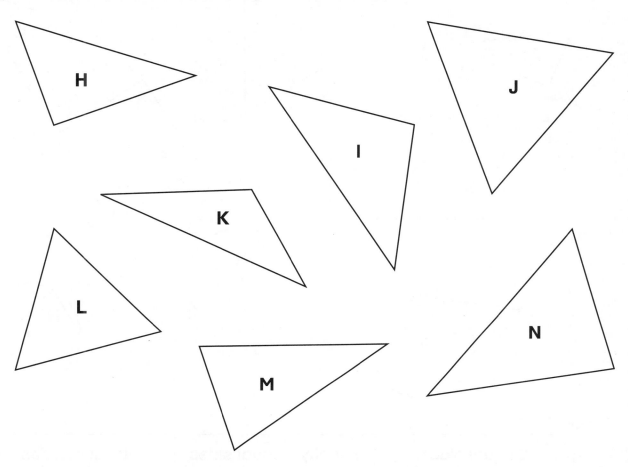

(a) Which triangles have 3 equal sides?

(b) Which triangles have only 2 equal sides?

(c) Which triangles have no equal sides?

(d) Which triangle has a right angle?

(e) Which triangles have an angle larger than a right angle?

Practice

2 All the circles below have the same radii.
Sort the labeled triangles according to the number of equal sides.

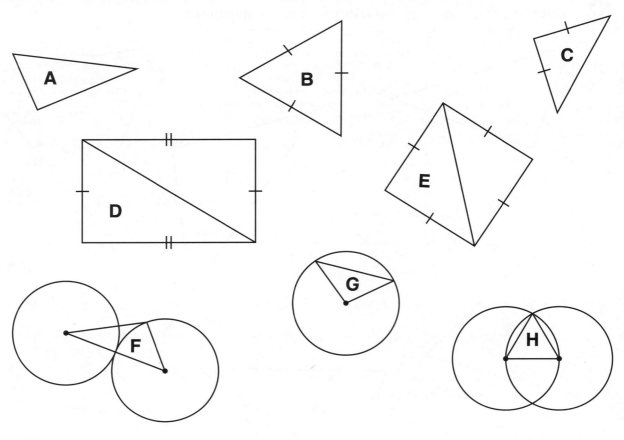

3 equal sides	Exactly 2 equal sides	0 equal sides

Challenge

3 Is it possible to draw a triangle that has sides of the following measurements?
Write "yes" or "no" next to each set of measurements.

(a) 5 cm, 4 cm, 11 cm

(b) 2 cm, 6 cm, 5 cm

(c) 11 cm, 6 cm, 7 cm

(d) 8 cm, 3 cm, 5 cm

Basics

1 Complete the sentences with 0, 1, 2, or 3.

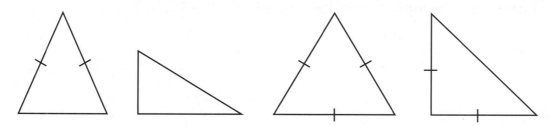

(a) A triangle can have _____, _____, or _____ equal angles.

(b) A triangle that has 2 equal sides has _____ equal angles.

(c) A triangle that has 0 equal sides has _____ equal angles.

(d) A triangle that has 3 equal sides has _____ equal angles.

(e) A triangle with 1 right angle can have _____ or _____ equal sides.

Practice

2 Measure the sides of each triangle in centimeters and write the number of equal angles.

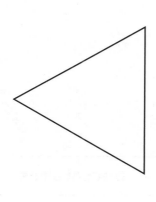

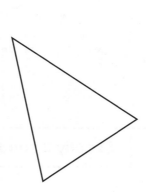

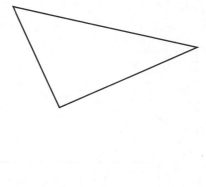

3 Use the circles and dots to help you draw three different triangles of each type.

All the circles are the same size so radii can be used to have sides of the same length.

Label your triangles, and then complete the table below.

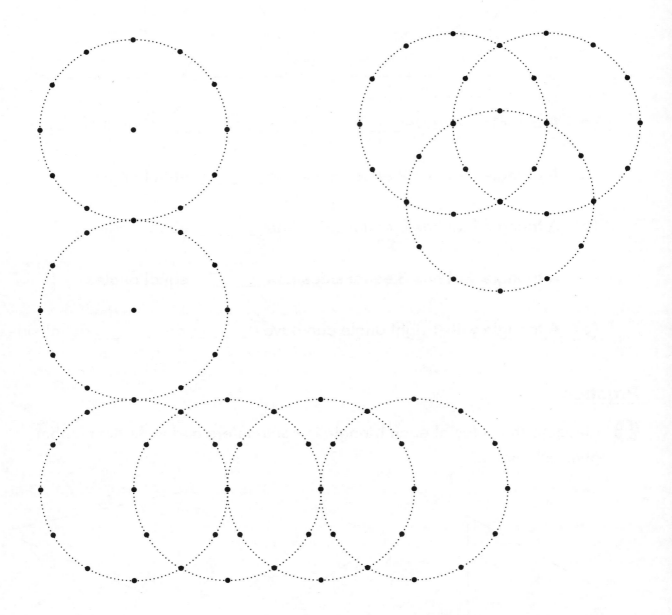

3 equal sides	Exactly 2 equal sides	0 equal sides

Challenge

4 The drawing shows how two toothpicks in a triangle made from six toothpicks can be moved to change one triangle into two smaller triangles.

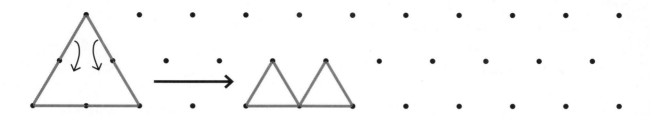

For each problem, all triangles should have three equal sides.

(a) Show how five toothpicks can be arranged to show two triangles.

(b) Show how three toothpicks can be removed to leave three triangles.

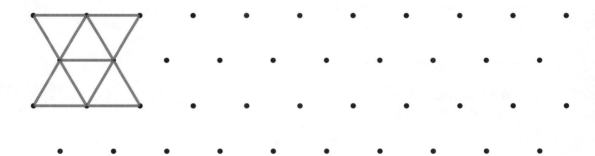

(c) Show how four toothpicks can be moved to show three triangles. The triangles do not have to be the same size.

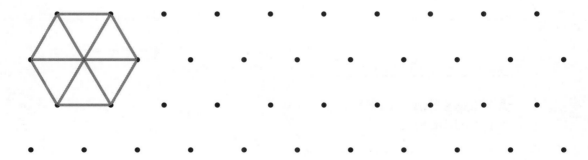

Basics

1 Match.

A rhombus

A quadrilateral with exactly one right angle

A quadrilateral with exactly two right angles

A rectangle that is not a square

A quadrilateral with no right angles

A shape that is not a quadrilateral

2 Color the rhombuses.
Draw a line to divide each quadrilateral into two triangles.

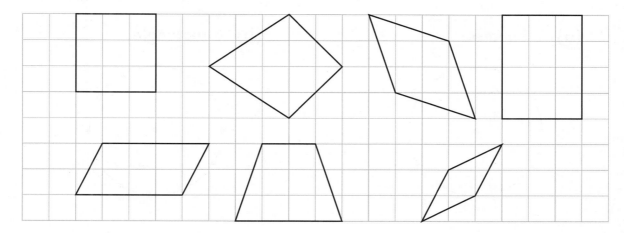

Practice

3 Use a ruler.
Draw a quadrilateral with...

(a) exactly one right angle.

(b) at least one angle
greater than a right angle.

(c) two angles smaller
than a right angle.

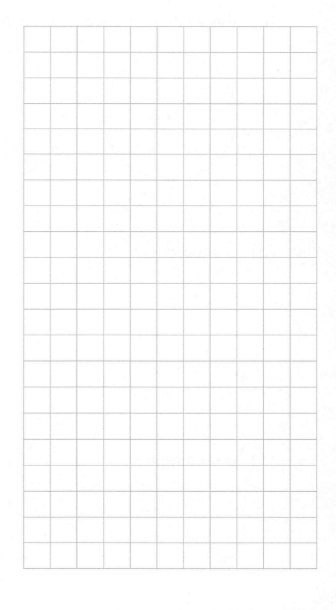

4 Use a set square or a ruler and a card with a right angle.
Connect four of the dots to make...

(a) a square.

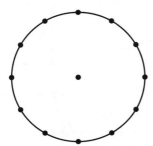

(b) a rectangle.

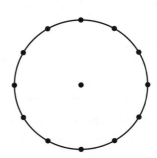

(c) three different quadrilaterals that are not rectangles.

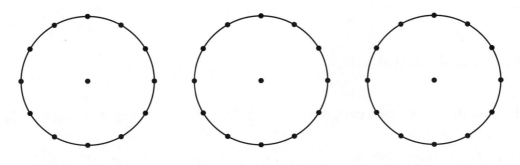

(d) a square.

(e) a rhombus.

(f) a rectangle.

Challenge

5 Four toothpicks can be arranged to show a rhombus:

(a) Show how ten toothpicks can be arranged to make three rhombuses.

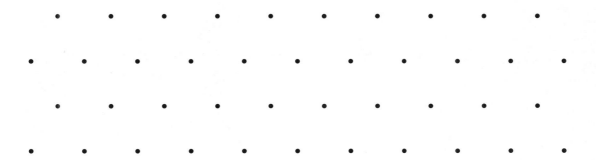

(b) Show how nine toothpicks can be arranged to make three rhombuses.

(c) Show how three toothpicks can be moved to make four rhombuses.

Basics

1. Use a compass and the centimeter graph below to draw a circle with a radius of...

   (a) 3 cm.

   (b) 5 cm.

2 One side of a triangle and the radii of circles with centers at each end of that side are given.
Draw the other two sides of each triangle with lengths of...

(a)   4 cm and 4 cm.

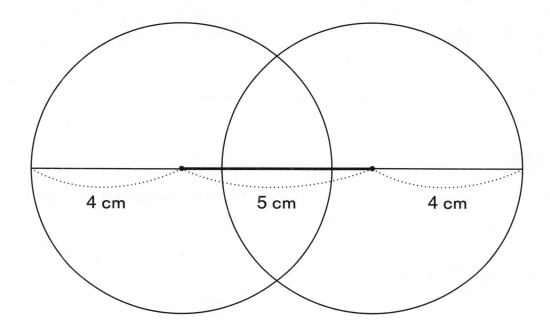

4 cm   5 cm   4 cm

(b)   3 cm and 2 cm.

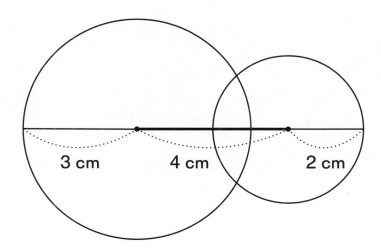

3 cm   4 cm   2 cm

3 The diagram below shows a line that is 4 cm long.

Parts of two circles with centers at the ends of the line and radii of 4 cm are shown.

Draw the other two sides of a triangle to create a triangle with 3 equal angles.

Practice

4 Use a compass and the centimeter graph below to draw a triangle with sides equal to...

(a)   9 cm, 6 cm, and 5 cm.

(b) 5 cm, 6 cm, and 6 cm.

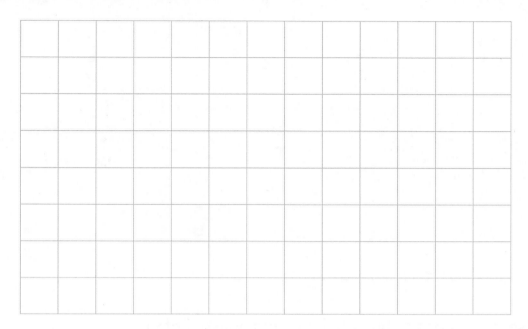

5 Draw a rhombus with sides of 5 cm, and a distance from one vertex to the opposite vertex of 3 cm.

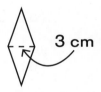

6

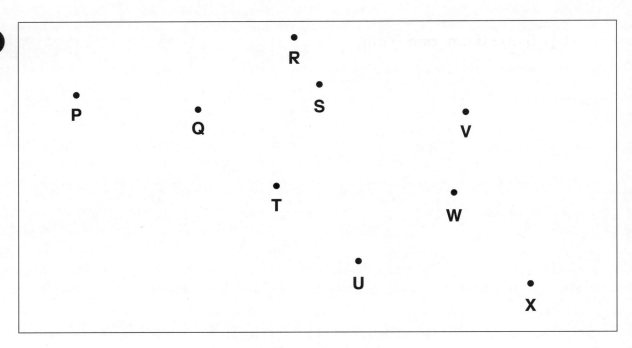

Use a compass to find all the points above that are...

(a) 3 cm from point T.

(b) 5 cm from point V.

(c) 6 cm from point R.

(d) 3 cm from point T and 5 cm from point V.

7 Use a compass to find the number of equal sides on the below triangles.

(a)

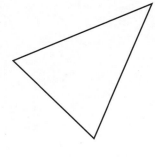

_____ equal sides

(b)

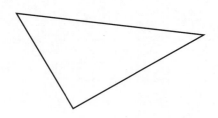

_____ equal sides

Check

1 A round satellite dish has a diameter of 28 cm.
What is its radius?

2 The rim of a basketball hoop has a diameter of 46 cm.
What is its radius?

3 Trace the two lines that form a right angle.

(a)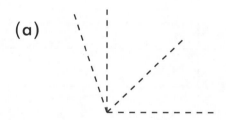

(b)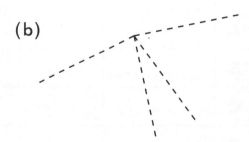

4 Write "true" or "false."

(a) All squares are rectangles.

(b) All rhombuses are squares.

(c) All quadrilaterals can be cut into two triangles.

(d) All triangles have at least one angle less than a right angle.

(e) All quadrilaterals have at least one angle less than a right angle.

5 (a) Complete the table below with the number of sides, angles, and types of angles.
Only consider inside angles.

	Sides	Angles	Angles smaller than a right angle	Right angles	Angles larger than a right angle
A					
B					
C					
D					
E					
F					
G					
H					

(b) List the shapes above that are quadrilaterals.

(c) List the shapes that are rhombuses.

6 Each diagram shows two circles of the same size and one smaller circle. The centers of the circles and their intersections are marked with dots. Connect the dots to form...

(a) a triangle with 3 equal angles.

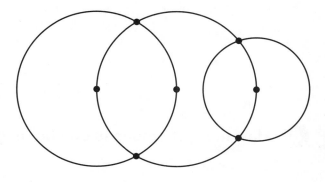

(b) a triangle with 2 equal angles.

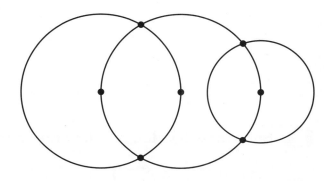

(c) a triangle with 0 equal angles.

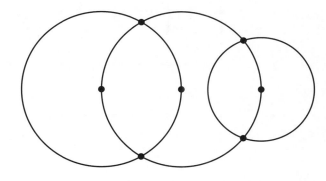

(d) a rhombus.

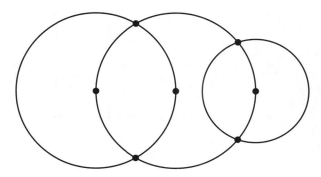

Challenge

7 How many rectangles can be made by joining the dots?

. . . . .

. . . . .

8 How many rhombuses can be made by joining the dots?

.

. .

.

. .

.

9 Show how two toothpicks can be removed to make two squares.

10 Show how two toothpicks can be moved to make four squares of the same size.

Chapter 13 Area and Perimeter

Basics

1 The area of ☐ is 1 square unit.

The area of ◺ is 1 half of a square unit.

(a) Find the area of each figure in square units.

J

_____ square units

K

_____ square units

L

_____ square units

M

_____ square units

N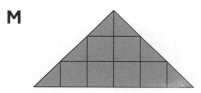

_____ square units

O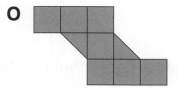

_____ square units

(b) Figure _____ has the largest area.

(c) Figure _____ has the smallest area.

(d) Figures _____ and _____ have the same area.

Practice

2 For each problem draw two different figures that have the same area as the one shown.

Then color them.

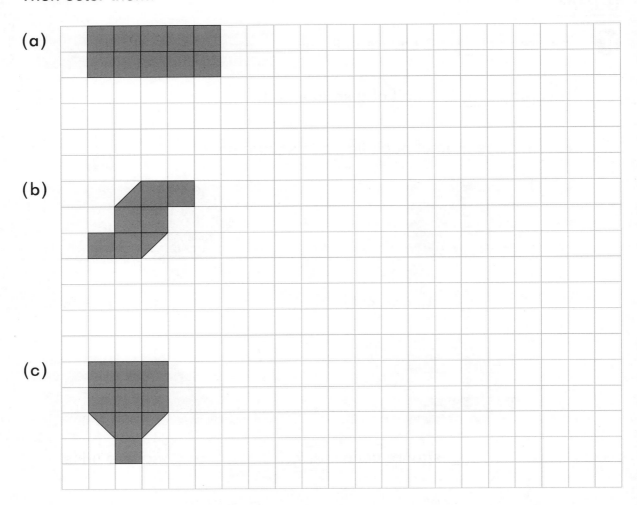

(a)

(b)

(c)

3 Draw another figure that has twice the area.

Then color it.

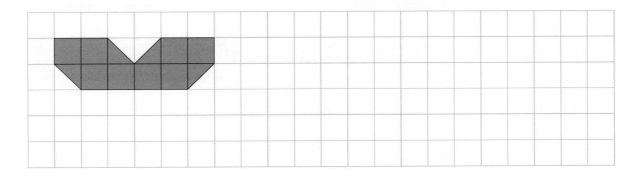

4 Draw another figure that has the same area as the given figure. Then color it.

5 Draw another figure that has three times the area as the given figure. Then color it.

Challenge

6 The first figure has an area of about 2 square units. What is the approximate area of the other 2 figures?

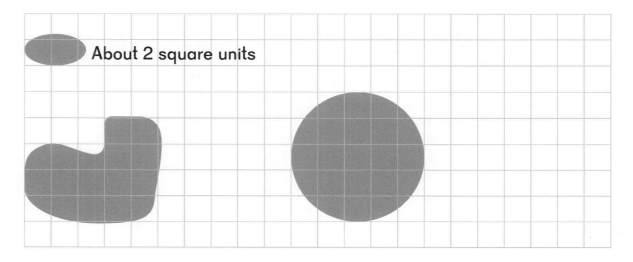

About _____ square units | About _____ square units

Basics

1 (a) Find the area of each figure in square centimeters.

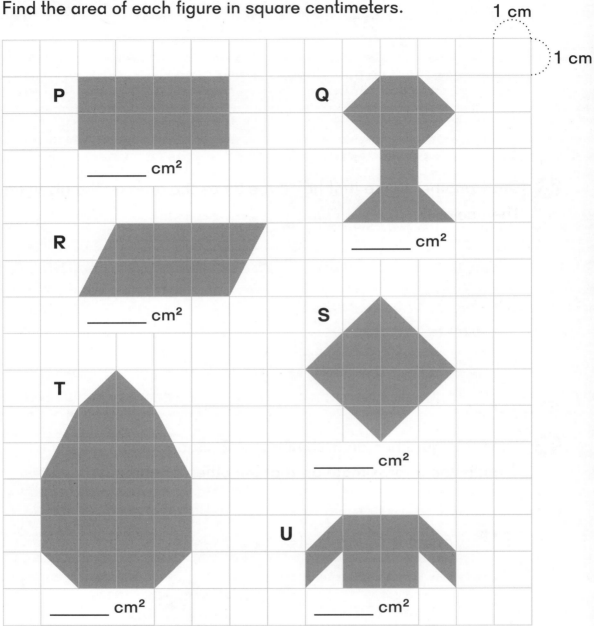

(b) Figure _____ has the largest area.

(c) Figure _____ has the smallest area.

(d) Which figures have the same area?

Practice

2 Each square on the grid below has an area of 1 cm².
Write the area of each figure, then draw another figure that is...

(a) the same area.

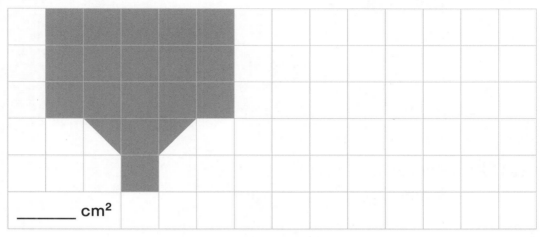

_____ cm²

(b) the same area.

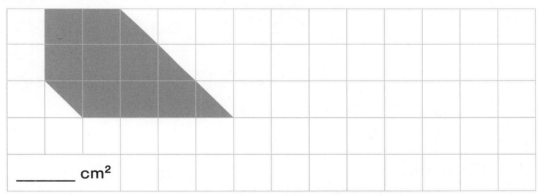

_____ cm²

(c) 3 cm² larger.

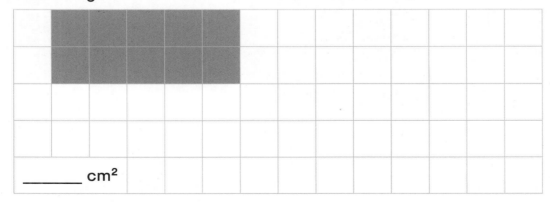

_____ cm²

(d) 6 cm² larger.

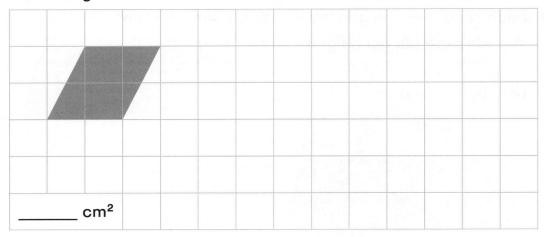

_____ cm²

(e) 2 cm² smaller.

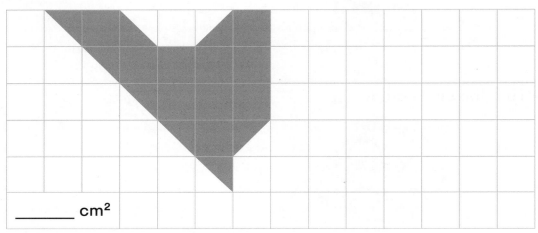

_____ cm²

(f) 5 cm² smaller.

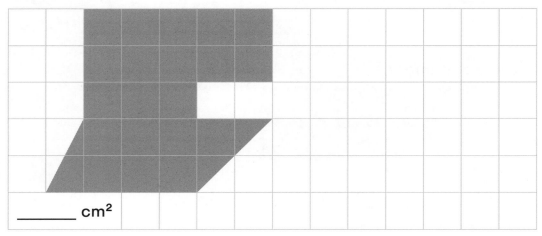

_____ cm²

3 Each square is 1 inch long.
The area of the shaded figure is _____.

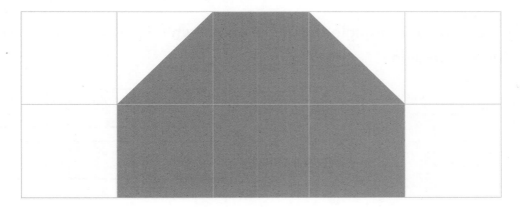

4 Fill in the blanks with in², m², or ft².

(a) The floor of a bedroom has an area of about 80 _____.

(b) A piece of paper has an area of about 94 _____.

(c) A car parking space is about 12 _____.

Challenge

5 Each square has an area of 1 cm².
What is the area of each triangle?

(a)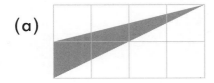

(b)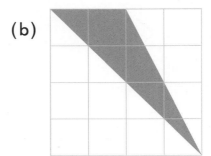

Basics

1 Find the area of each rectangle in square units by multiplying the number of units along each side.

(a)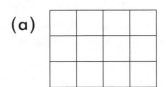

Length × Width = [   ] × [   ] = [   ]

Width × Length = [   ] × [   ] = [   ]

Area: _____ square units

(b)

[   ] × [   ] = [   ]

Area: _____ square units

(c)

Area: _____ square units

(d)

Area: _____ square units

2 Find the area of each rectangle in cm².

Area: _____ cm²

Area: _____ cm²

Area: _____ cm²

Area: _____ cm²

1 cm

1 cm

Practice

3 Measure the length of the sides of each rectangle in centimeters. What is the area of each rectangle?

(a)

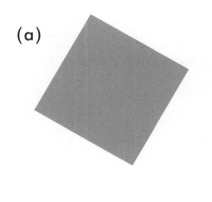

Area: _____

(b)

Area: _____

4 What is the area of rectangles with the given lengths for the sides?

(a) 15 m, 9 m

(b) 12 in, 7 in

Challenge

5 1-centimeter tiles are used to make rectangles with the following areas. What are the possible lengths of the sides?

(a) 24 cm²

(b) 27 cm²

6 This rectangle is made up of square tiles that are each 2 cm long. What is the area of the rectangle?

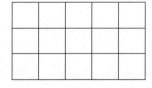

7 This rectangle is made up of tiles that are 3 cm long and 2 cm wide. What is the area of the rectangle?

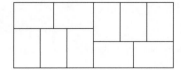

Basics

1 (a) A rectangle and a square are combined along their edges to form a new figure.
What is the area of the new figure?

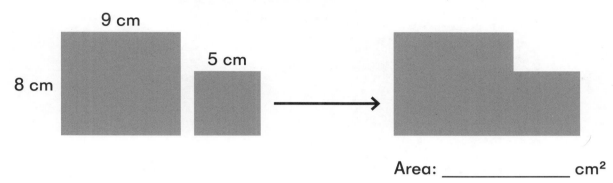

Area: _____ cm²

(b) A square is cut out of a rectangle as shown to form a new figure.
What is the area of the new the figure?

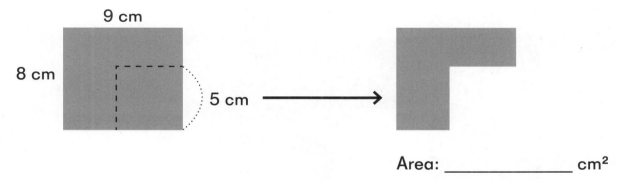

Area: _____ cm²

(c) Three rectangles of the same size are combined along their edges to form a new figure.
What is the area of the new figure?

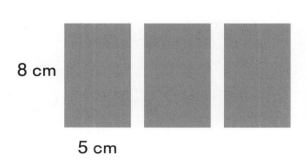

Area: _____ cm²

2 Measure the sides of these figures in centimeters.
Find the area of each figure.

(a)

Area: _____

(b)

Area: _____

(c)

Area: _____

Practice

3 A square and a rectangle of the lengths shown are combined.
Complete the equations to show two ways of finding the area.

8 cm

5 cm   8 cm

(a) 8 × 5 = 40

8 × 8 = [   ]

40 + [   ] = [   ]

(b) 8 + 5 = [   ]

8 × [   ] = [   ]

The area is _____ cm².

4 The numbers indicate the lengths in equal units.
Find the shaded areas of each figure in square units.

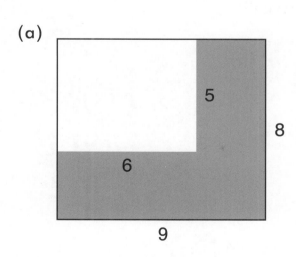

(a)

5

8

6

9

(b)

8

2   2

2   2

8

2   2

2   2

Area: _____

Area: _____

5 The numbers indicate the length of the sides in equal units.
Find the shaded area of the figure in square units.

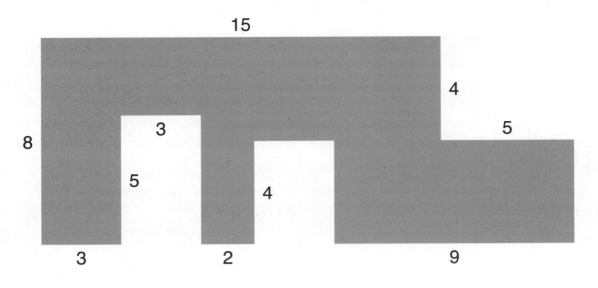

Area: _____

Challenge

6 The numbers indicate the length of the sides in units.
Find the area of each figure in square units.

(a)

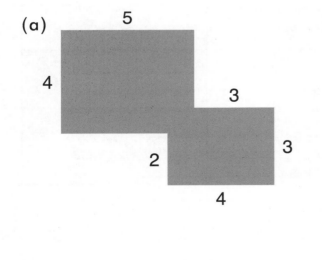

Area: _____

(b)

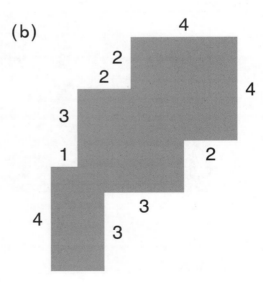

Area: _____

Check

1 Draw a figure with an area that is 7 square units more than the area of the given figure.

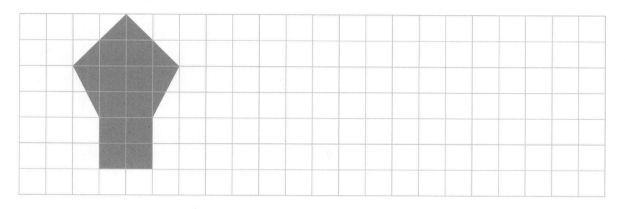

2 Draw two different figures with areas of...

(a) 17 square units.

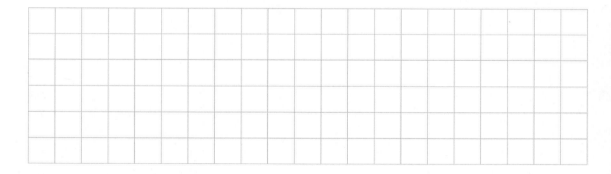

(b) 23 square units.

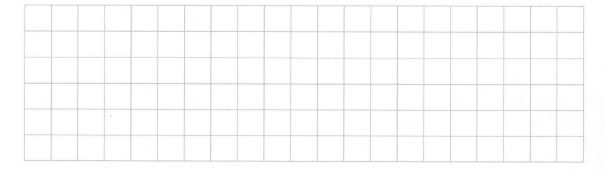

3 Three rectangles are 2 cm by 2 cm, 4 cm by 3 cm, and 5 cm by 3 cm.
Draw a figure showing the 3 rectangles joined along the edges with no overlap.
What is the area of the figure?

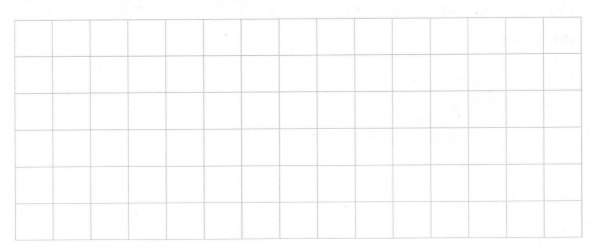

4 Draw two different rectangles with areas of 16 cm².

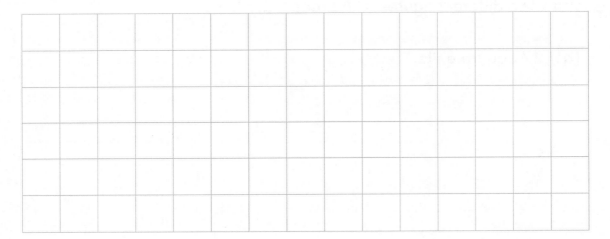

5 A rectangular sign is 32 cm long and 8 cm wide.
What is the area of the sign?

6 A rectangular piece of plywood is 4 ft wide.

It is 3 times as long as it is wide.

What is the area of the piece of plywood?

7 A piece of cardboard is 54 cm long.

It is 6 times as long as it is wide.

What is the area of the piece of cardboard?

8 A square piece of paper has an area of 36 cm².

A smaller square measuring 3 cm on one side is cut from the paper.

What is the area of the remaining piece of paper?

9 The cost of installing a certain carpet is $8 per square foot.

How much would it cost to carpet a room with the following dimensions?

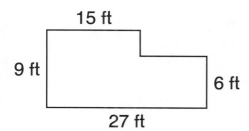

15 ft

9 ft

6 ft

27 ft

Challenge

10 Two squares with side lengths of 10 cm overlap by 4 cm.
What is the area of the resulting rectangle?

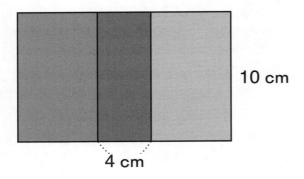

10 cm

4 cm

11 By how much should the following rectangles overlap in order to create
a rectangle with an area of 144 cm²?

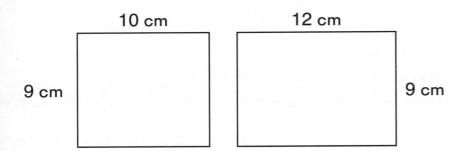

10 cm

12 cm

9 cm

9 cm

12 Plastic rectangular tiles 2 cm wide and 3 cm long are being used to cover
a piece of cardboard that is 10 cm by 13 cm.
What is the greatest number of tiles that can be used?

Basics

1 A triangle has sides that are 5 cm, 6 cm, and 4 cm long.
Find the perimeter of the triangle.

5 + 6 + 4 = ⬚

Perimeter: _____ cm

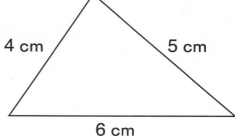

4 cm   5 cm

6 cm

2 Measure the sides of these figures in centimeters and find their perimeters.

(a)

Perimeter: _____

(b)

Perimeter: _____

(c)

Perimeter: _____

Practice

3 The lengths of the sides of each figure are marked in equal units.
Find the perimeter of each figure in units.

(a)

6

3   3

6

Perimeter: _____

(b)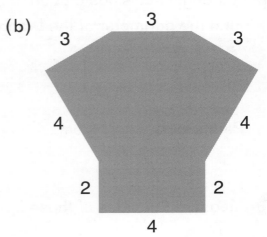

Perimeter: _____

(c)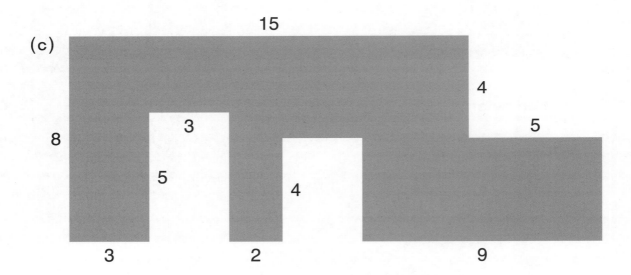

Perimeter: _____

4 A hexagon has 6 sides, each with a length of 12 cm.
What is the perimeter of the hexagon?

5 The sum of the lengths of 3 sides of a rhombus is 84 cm.
What is the perimeter of the rhombus?

Challenge

6 The numbers indicate the length of the sides in units.
Find the perimeter of each figure.

(a)

(b)

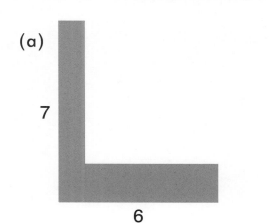

7

6

4

5

3

2

Perimeter: _____

Perimeter: _____

Basics

1 A rectangle is 9 cm long and 5 cm wide.
Complete the equations to show two different methods to calculate the perimeter.

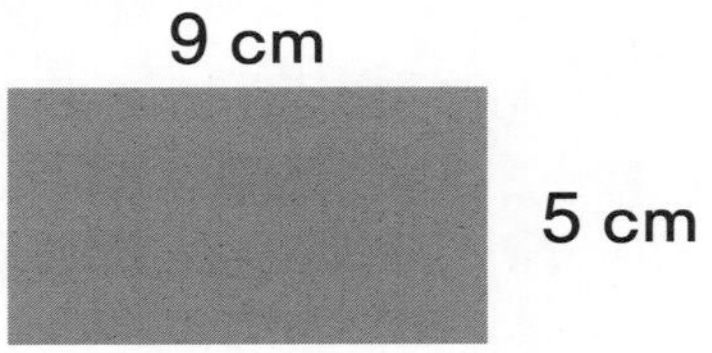

9 cm

5 cm

(a) 9 + 5 = 14

14 × 2 = ☐

(b) 9 × 2 = 18

5 × 2 = ☐

18 + ☐ = ☐

The perimeter is _____ cm.

2 One side of a square is 7 cm long.
Complete the equation to find the perimeter.

7 cm

4 × ☐ = ☐

The perimeter is _____ cm.

Practice

3 Find the perimeters of the following rectangles in units.
The lengths of the sides are marked in equal units.

(a)

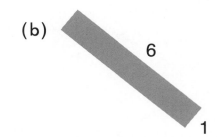

3

5

Perimeter: _____

(b)

6

1

Perimeter: _____

4 Find the perimeters of the following squares in units.
The lengths of a side is marked in equal units.

(a)

6

Perimeter: _____

(b)

4

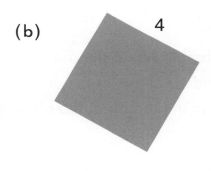

Perimeter: _____

5 A rectangular sign is 15 m long and 8 m wide.
What is the perimeter of the sign?

6 A square piece of paper has a side of 18 cm.
What is its perimeter?

7 A square has a perimeter of 40 cm.
What is the length of one side?

Challenge

8 A rectangle has a perimeter of 30 cm.
One side is 8 cm long.
What is the length of the other side?

9 A square has a perimeter of 32 cm.
It is cut into four equal squares as shown.
What is the perimeter of one of the small squares?

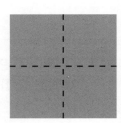

Basics

1 Each small square on the grid has a side length of 1 unit.

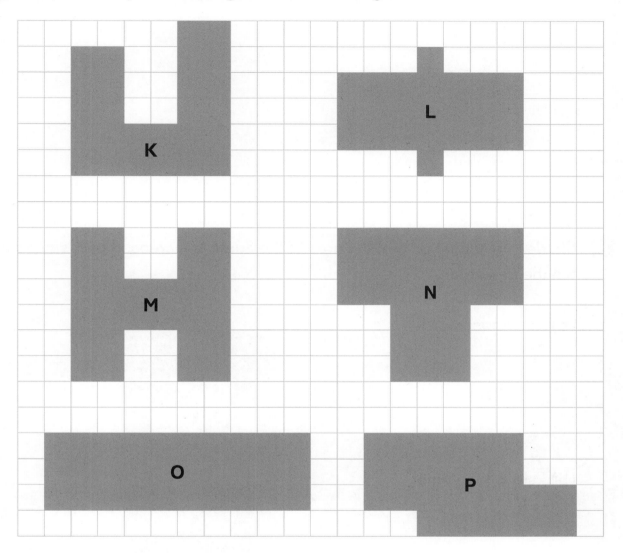

(a) Which figures have the same area and different perimeters?

(b) Which figures have the same perimeter but different areas?

(c) Which figures have the same area and perimeter?

Practice

2 Use the grid to draw three different rectangles with perimeters of 16 units.
Write the area of each figure.
Which figure has the smallest area?

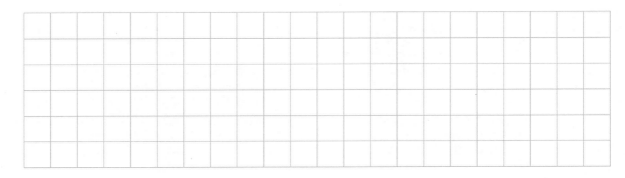

3 Shade one square so that the area of the figure is increased by 1 square unit, but the perimeter...

(a) decreases.

(b) stays the same.

(c) increases.

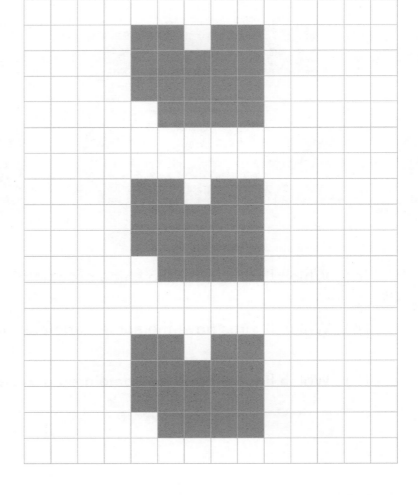

4 Draw another figure that has...

(a) the same area and perimeter.

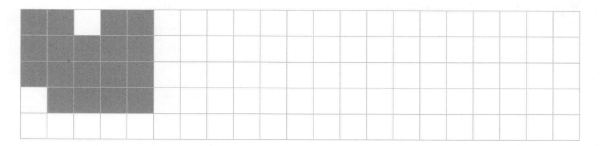

(b) a greater perimeter but smaller area.

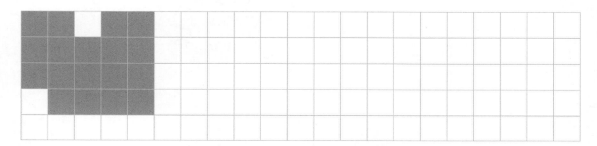

Challenge

5 Draw a rectangle with the largest possible perimeter that has a smaller area than the given figure.
The sides of the rectangle must be whole numbers of units.

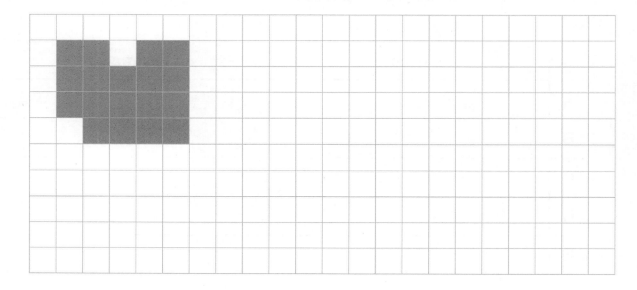

Check

1 The following figures are made up of square units.
The numbers indicate the length of the sides in units.
Find the area and perimeter of each figure.

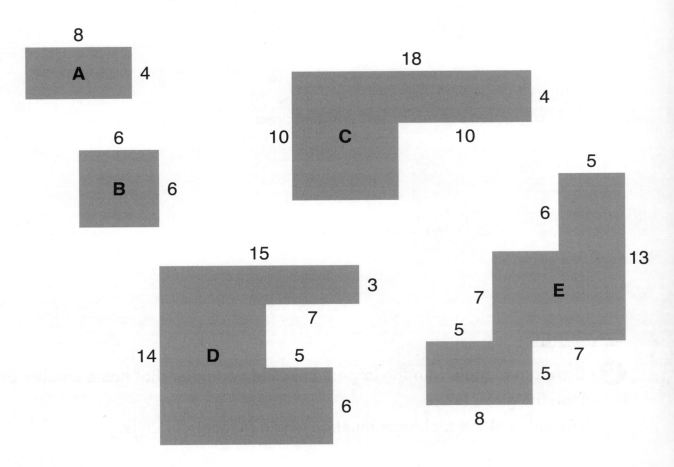

	Area	Perimeter
A		
B		
C		
D		
E		

2 A rectangle has a length of 16 cm and a width of 9 cm.
What is its area and perimeter?

3 The diameter of each circle is 6 units.
What is the perimeter of the quadrilateral?

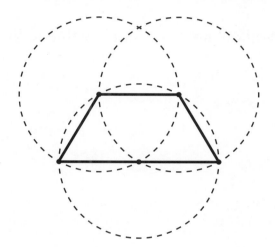

4 Draw a figure with an area of 35 square units and a perimeter of 26 units.

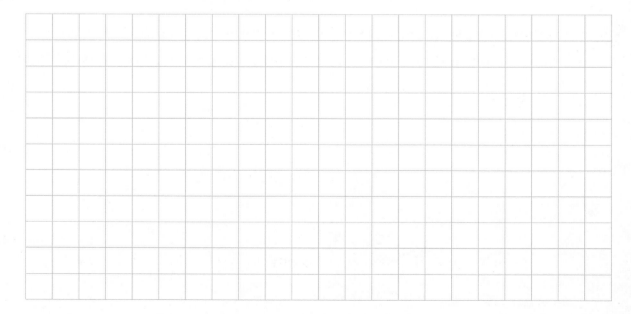

Challenge

5 A square is cut into four equal squares as shown.
The perimeter of each smaller square is 12 cm.
What is the area of the larger square?

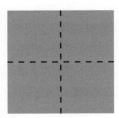

6 The figure below is made up of three identical squares, each with a side of 6 cm.
What is the perimeter of the figure?

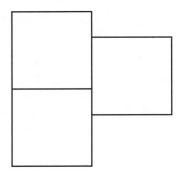

7 A rectangular piece of paper is folded to form the shape shown below.
What is the perimeter and area of the piece of paper before it is folded?

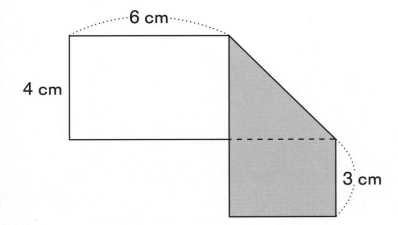

4 cm

6 cm

3 cm

Chapter 14 Time

Basics

1 Fill in the blanks with min for minutes, s for seconds, or h for hours.

(a) You should wash your hands for at least 20 _____.

(b) You should brush your teeth for at least 2 _____.

(c) Colton's piano recital lasted for 2 _____.

(d) Aliya ran halfway around the football field in about 45 _____.

(e) A soccer game has two 45-_____ halves.

(f) It takes about 6 _____ to fly by airplane across the United States.

2 Write a.m. or p.m. in the blanks.

(a) 10:00 _____ is between midnight and noon.

(b) Jamal went to bed for the night at 9:00 _____.

(c) 5 hours after 3:00 a.m. is 8:00 _____.

(d) 4 hours before 3:00 a.m. is 11:00 _____.

(e) The restaurant is open from 6:30 _____ to 10:00 _____.

3 (a) 1 h = 60 min

3 h = [ ] min

3 h 15 min = [ ] min

(b) 60 min = [ ] h

120 min = [ ] h

130 min = [ ] h [ ] min

4 (a) 1 min = 60 s

2 min = [ ] s

2 min 45 s = [ ] s

(b) 60 s = [ ] min

180 s = [ ] min

200 s = [ ] min [ ] s

5 (a) 1 day = [ ] h

5 days = [ ] h

(b) 1 week = [ ] days

15 weeks = [ ] days

Practice

6 Complete the tables.

Hours	Minutes
1	
2	
3	
4	
5	

Hours	Minutes
6	
7	
8	
9	
10	

7 Match.

1 h 5 min	230 min
2 h 25 min	105 min
1 h 45 min	65 min
4 h 15 min	80 min
3 h 50 min	255 min
2 h 35 min	155 min
1 h 20 min	145 min

8 (a) 5 min 5 s = ☐ s

(b) 1 min 32 s = ☐ s

(c) 140 s = ☐ min ☐ s

(d) 335 s = ☐ min ☐ s

9 (a) 1 week 2 days = ☐ days

(b) 16 days = ☐ weeks ☐ days

10 Dexter went away on a trip that lasted 3 weeks and 4 days.
How many days was he gone?

11 Valentina practiced the bass for 85 minutes.
Onowa practiced the piano for 1 hour 20 minutes.
Who practiced longer and how much longer?

12 Caden ran a marathon in 228 minutes.
Diego ran the same marathon in 4 hours 13 minutes.
Who ran the marathon faster and how much faster?

Challenge

13 366 days = ☐ weeks ☐ days

14 How many days are in a month that begins and ends on a Wednesday?

Basics

1 Complete the following:

(a)

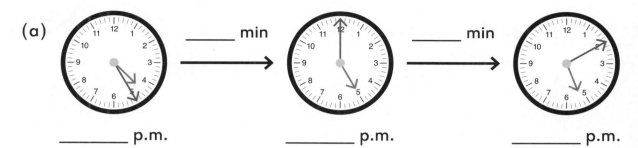

_____ min   _____ min

_____ p.m.   _____ p.m.   _____ p.m.

45 minutes after 4:25 p.m. is _____ p.m.

_____

(b)

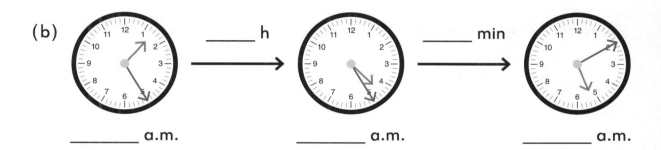

_____ h   _____ min

_____ a.m.   _____ a.m.   _____ a.m.

3 hours and 45 minutes after 1:25 a.m. is _____ a.m.

_____

(c) 1 h 25 min + 3 h 45 min = ☐ h ☐ min

2

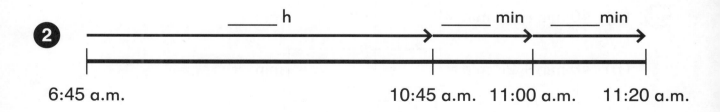

_____ h   _____ min   _____ min

6:45 a.m.   10:45 a.m. 11:00 a.m.   11:20 a.m.

_____ hours and _____ minutes pass from 6:45 a.m. to 11:20 a.m.

Practice

3 Write the amount of time that passes from...

(a) 4:35 a.m. to 7:00 a.m.

(b) 4:35 a.m. to 7:20 a.m.

(c) 4:35 a.m. to 10:20 a.m.

4 What time is it...

(a) 7 hours after 1:10 p.m.?

(b) 7 hours 50 minutes after 1:10 p.m.?

(c) 7 hours 55 minutes after 1:10 p.m.?

5 (a) 15 min + ☐ min = 1 h   (b) 20 min + ☐ min = 1 h

(c) 35 min + ☐ min = 1 h   (d) 56 min + ☐ min = 1 h

(e) 47 min + ☐ min = 1 h   (f) 9 min + ☐ min = 1 h

6 (a) 40 min + 25 min = ☐ h ☐ min

(b) 40 min + 28 min = ☐ h ☐ min

(c) 32 min + 32 min = ☐ h ☐ min

(d) 25 min + 49 min = ☐ h ☐ min

7 A movie started at 7:35 p.m.
It lasted 2 hours and 43 minutes.
What time did the movie end?

8 Ricardo got to the airport at 6:20 a.m.
His flight left at 8:10 a.m.
How long was he at the airport?

9 Laura spent 1 hour 20 minutes mowing the lawn and another
50 minutes trimming the grass along the walk.
How long did she spend on both tasks altogether?
Give your answer in compound units.

Challenge

10 What time is 165 minutes after 4:30 p.m.?

Check

1 This chart shows the time it took for some girls to swim 200 meters.

Renata	165 s
Taylor	2 min 15 s
Natalia	190 s
Violet	3 min 5 s

Write the times in order from shortest time to longest time.

2 (a) 3 h 40 min + 4 h 50 min = ⬚ h ⬚ min

(b) 5 h 50 min + 2 h 43 min = ⬚ h ⬚ min

(c) 11 h 17 min + 6 h 25 min = ⬚ h ⬚ min

(d) 1 h 10 min + 2 h 25 min + 3 h 45 min = ⬚ h ⬚ min

3 Write the amount of time that passes from...

(a) 2:35 p.m. to 4:00 p.m.

(b) 8:50 a.m. to 11:39 a.m.

(c) 1:22 p.m. to 3:40 p.m.

4 If yesterday was Tuesday, what day of the week is it 1 week and 4 days from today?

5 Kim took 62 days to hike the Pacific Northwest Trail. Write this time in weeks and days.

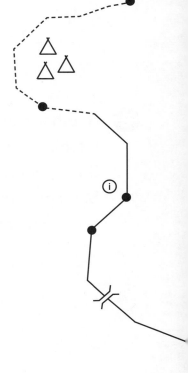

6 A show began at 4:35 p.m. and ended at 6:20 p.m. How long did the show last?

7 Megan started exercising at 6:12 a.m. She exercised for 1 hour 42 minutes. What time did she finish exercising?

8 Isaac left home at 8:50 a.m.

He took 33 minutes to travel to the park.

He spent 2 hours 20 minutes at the park.

What time did he leave the park?

9 Heather started an art project at 4:30 p.m.

It took her 15 minutes to gather the material and set up her canvas.

She spent 2 hours 20 minutes painting.

She spent another 25 minutes cleaning up.

What time did she finish?

Challenge

10 A bus service runs every 15 minutes.

There is a bus at 9:10 a.m.

How many buses are there between 9:00 a.m. and 11:00 a.m.?

Basics

1 Write the times using a.m. or p.m.

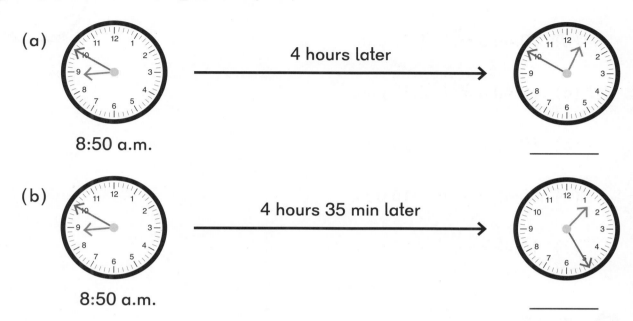

(a) 8:50 a.m.   4 hours later   _____

(b) 8:50 a.m.   4 hours 35 min later   _____

(c) 6 hours and 35 minutes after 8:50 a.m. is _____.

2 (a) 8:45 p.m. to midnight is _____ hours and _____ minutes.

(b) Midnight to 6:30 a.m. is _____ hours and _____ minutes.

(c) 8:45 p.m. to 6:30 a.m. is _____ hours and _____ minutes.

3 Complete the following:

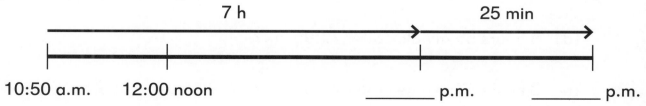

7 h   25 min

10:50 a.m.   12:00 noon   _____ p.m.   _____ p.m.

7 h 25 minutes after 10:50 a.m. is _____.

Practice

4 Write the amount of time that passes from...

(a) 10:40 a.m. to 12:00 noon.

(b) 10:40 a.m. to 1:40 p.m.

(c) 10:40 a.m. to 2:15 p.m.

(d) 10:40 a.m. to 2:18 p.m.

(e) 10:42 a.m. to 2:15 p.m.

(f) 10:42 a.m. to 2:18 p.m.

5 What time is it...

(a) 8 hours 50 minutes after 7:25 p.m.?

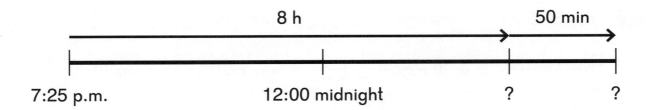

(b) 8 hours 55 minutes after 7:25 p.m.?

(c) 8 hours 57 minutes after 7:25 p.m.?

(d) 8 hours 55 minutes after 7:29 p.m.?

(e) 8 hours 57 minutes after 7:29 p.m.?

6 A store opens at 6:30 a.m. and closes at 10:45 p.m.

(a) How long is it open during the day?

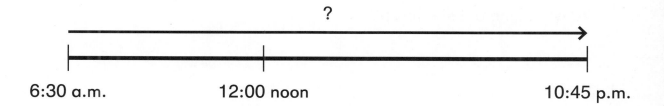

6:30 a.m.   12:00 noon   10:45 p.m.

(b) How long is it closed during the night?

7 Mr. Ikeda left home for work at 7:35 a.m.
He arrived back home 9 hours and 45 minutes later.
What time did he arrive back home?

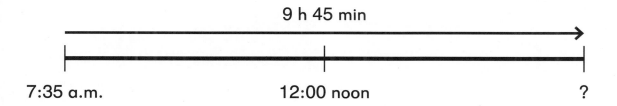

7:35 a.m.   12:00 noon   ?

8 Sasha went to bed at 8:50 p.m.
She read for 25 minutes and fell asleep 13 minutes after she finished reading.
She slept for 8 hours and 10 minutes.
What time did she wake up?

9 Laura started working in the yard at 11:25 a.m.
She weeded her garden for 1 hour and 30 minutes.
She trimmed her fruit trees for 2 hours and 15 minutes.
Then she spent 25 minutes tidying up and putting away her tools.
At what time did she finish?

Challenge

10 How many hours and minutes are there between 9:45 p.m. on Friday and 6:15 a.m. on Sunday?

11 A bus service runs every 25 minutes.
There is a bus at 9:10 a.m.
How many buses are there between 9:00 a.m. and 4:00 p.m.?

Basics

1 (a) 3 h − 45 min = 2 h [   ] min

      2 h   60 min

(b)   3 h 10 min − 45 min = 2 h [   ] min

    2 h 10 min 60 min

(c)   7 h 10 min $\xrightarrow{-4\ h}$ 3 h 10 min $\xrightarrow{-45\ min}$ [   ] h [   ] min

(d)   4 hours 45 minutes before 7:10 a.m. is _____.

2 (a) 3 h − 10 min = 2 h [   ] min

(b)   4 h 25 min − 35 min = [   ] h [   ] min

      25 min 10 min

(c)   4 h 25 min − 1 h 35 min = [   ] h [   ] min

(d)   1 hour 35 minutes before 4:25 p.m. is _____.

3 (a) 54 min = 1 h − [   ] min

(b)   3 h 20 min $\xrightarrow{-1\ h}$ 2 h 20 min $\xrightarrow{+6\ min}$ [   ] h [   ] min

(c)   5 h 20 min − 2 h 54 min = [   ] h [   ] min

(d)   2 hours 54 minutes before 5:20 p.m. is _____.

Practice

4 What time is it...

(a) 35 minutes before 4:35 p.m.?

(b) 45 minutes before 4:35 p.m.?

(c) 1 hour and 45 minutes before 4:35 p.m.?

(d) 1 hour and 48 minutes before 4:35 p.m.?

5 (a) 5 h – 32 min = ☐ h ☐ min

(b) 5 h 10 min – 32 min = ☐ h ☐ min

6 (a) 5 h 55 min – 2 h 24 min = ☐ h ☐ min

(b) 7 h 32 min – 3 h 50 min = ☐ h ☐ min

(c) 20 h 5 min – 18 h 30 min = ☐ h ☐ min

7 What time is it...

(a) 8 hours 25 minutes before 10:05 p.m.?

(b) 3 hours 55 minutes before 11:35 a.m.?

8 Mia spent 3 hours 45 minutes at a park.
She left the park at 4:35 p.m.
What time did she get to the park?

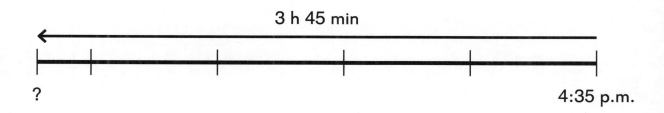

3 h 45 min

?   4:35 p.m.

9 Megan exercised for 1 hour 25 minutes on Monday.
She exercised for 2 hours 15 minutes on Wednesday.
How much longer did she exercise on Wednesday than on Monday?

10 Arman spent 5 hours 15 minutes at the lake.
He spent 35 minutes having a picnic, 1 hour 20 minutes swimming,
and the rest of the time fishing.
How much time did he spend fishing?

11 A concert will start at 7:30 p.m.

Ethan wants to reach the performance hall 15 minutes before the start time.

It will take him 1 hour 25 minutes to get to the hall from his house.

What time should he leave his house?

Challenge

12 A clock loses 3 minutes every hour.

If it was set to the correct time at 10:00 a.m. on Friday, what time will it show at 10:00 a.m. on Saturday?

13 A bus service runs every 20 minutes.

There is a bus at 9:10 a.m.

How many buses are there between 6:00 a.m. and noon?

Exercise 6

Basics

1 Complete the following.

(a)

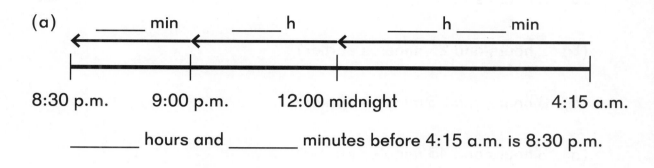

_____ min   _____ h   _____ h _____ min

8:30 p.m.   9:00 p.m.   12:00 midnight   4:15 a.m.

_____ hours and _____ minutes before 4:15 a.m. is 8:30 p.m.

(b)

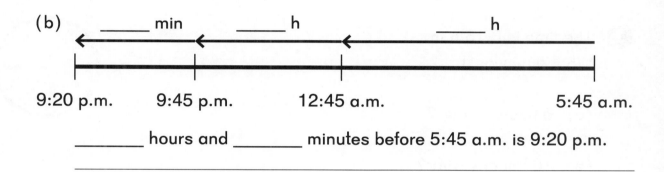

_____ min   _____ h   _____ h

9:20 p.m.   9:45 p.m.   12:45 a.m.   5:45 a.m.

_____ hours and _____ minutes before 5:45 a.m. is 9:20 p.m.

(c)

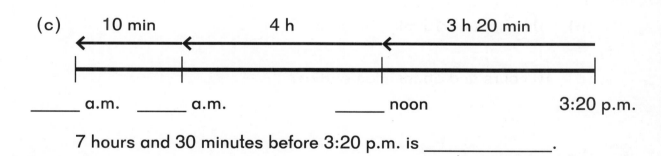

10 min   4 h   3 h 20 min

_____ a.m.   _____ a.m.   _____ noon   3:20 p.m.

7 hours and 30 minutes before 3:20 p.m. is _____.

(d)

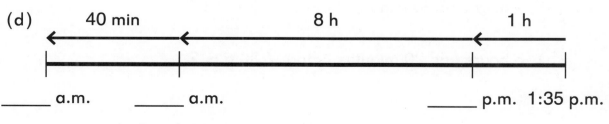

40 min   8 h   1 h

_____ a.m.   _____ a.m.   _____ p.m.   1:35 p.m.

9 hours and 40 minutes before 1:35 p.m. is _____.

Practice

2 The time is 2:25 p.m.
What time was it...

(a) 2 hours earlier?

(b) 2 hours and 25 minutes earlier?

(c) 3 hours and 25 minutes earlier?

(d) 3 hours and 45 minutes earlier?

3 The time is 6:15 a.m.
What time was it...

(a) 6 hours earlier?

(b) 10 hours earlier?

(c) 10 hours and 15 minutes earlier?

(d) 10 hours and 35 minutes earlier?

4 What time was it...

(a) 1 hour and 20 minutes before 12:30 p.m.?

(b) 8 hours and 20 minutes before 7:30 p.m.?

5 Austin arrived home at 2:35 p.m.

He had been gone for 3 hours and 30 minutes.

What time did he leave home?

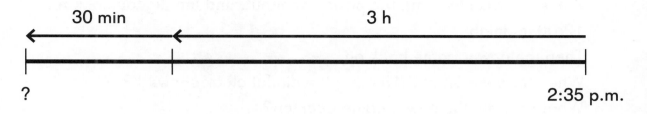

30 min   3 h

?   2:35 p.m.

6 A store closes at 10:30 p.m.

It was open for 15 hours and 15 minutes.

What time did it open?

7 Sunrise was at 7:05 a.m.

It set 11 hours and 10 minutes earlier.

What time was sunset the night before?

Challenge

8 Franco has an analog clock that runs on batteries and a digital clock that runs on electricity.

While he was sleeping, the power went out, and the digital clock reset to 12:00 midnight.

Then the power came back on.

When he woke up at 6:15 a.m., the digital clock showed 2:37 a.m.

What time did the power come back on?

9 A bus service runs every 35 minutes.

There is a bus at 2:10 p.m.

How many buses are there between 11:00 a.m. and 4:00 p.m.?

Check

1 (a) 7 h 10 min − 5 h 25 min = ☐ h ☐ min

(b) 3 h 55 min − 2 h 24 min = ☐ h ☐ min

(c) 3 h 25 min + ☐ h ☐ min = 7 h 35 min

(d) 7 h 50 min + ☐ h ☐ min = 10 h 15 min

(e) ☐ h ☐ min + 12 h 30 min = 14 h 7 min

2 Sunset was at 8:45 p.m.
The sun rose 8 hours and 45 minutes later.
What time was sunrise?

3 Papina practiced the violin for 1 hour 20 minutes.
She then played outside with her friends for 3 hours 45 minutes.
How much more time did she spend playing than practicing?

4 Colton started playing tennis at 11:45 a.m. and finished at 1:15 p.m. How many minutes did he play tennis?

5 A program ended at 12:45 p.m.
It was 2 hours and 30 minutes long.
When did it start?

6 A restaurant opens every day at 11:00 a.m. for lunch, closes at 2:45 p.m., opens again at 5:30 p.m. for dinner, and closes at 11:30 p.m.
How long is it open in one day?

7 Misha did a project on millipedes for school.
She spent 1 hour 35 minutes researching the topic and another 45 minutes writing the report.
She finished her report at 5:10 p.m.
What time did she start her project?

Challenge

8 What time will it be 800 minutes after 10:30 a.m.?

9 Sydney has an analog clock and a digital clock that both run on electricity. When the power goes out, both stop running, but only the digital clock resets to 12:00 midnight.

(a) One night, the power went off at 3:00 a.m. for 25 minutes. At 6:15 a.m., what time will show on each clock?

(b) Another night, Sydney woke up to find her analog clock displaying 5:05 and her digital clock displaying 2:45 a.m. What time was it when the power went out?

10 There are 365 days in a year, except for a leap year, which has 366 days.
The year 2000 was a leap year, and every 4 years after that.
The chart shows the number of days in each month in a non-leap year.

Jan.	Feb.	Mar.	Apr.	May	June	July	Aug.	Sept.	Oct.	Nov.	Dec.
31	28	31	30	31	30	31	31	30	31	30	31

Every leap year, February has 29 days instead of 28.

Write down your birth date and today's date.
How many days old are you?

Chapter 15 Money

Basics

1 (a) Write the amount of money each person has in dollars and cents and in cents only.

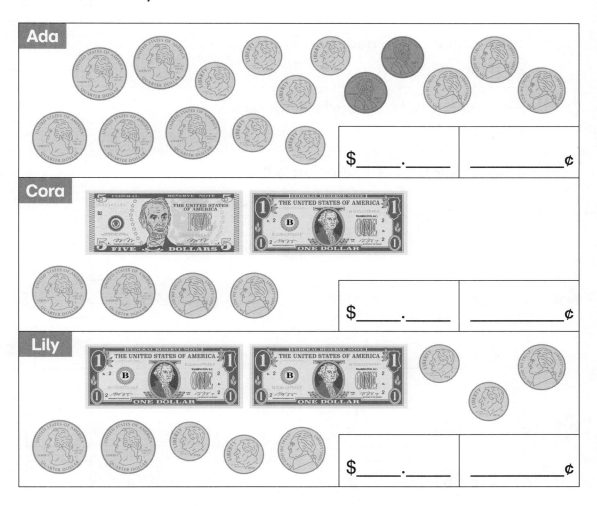

(b) Ada needs $_____ to have the same amount of money as Lily.

(c) If Ada gets 5 more quarters, she will have $_____.

(d) _____ has the greatest amount of money.

Practice

2 Write the amount of money in dollars and cents and in cents only.

2 quarters, 2 dimes, 2 nickels, 2 pennies	$_____	_____ ¢
3 quarters, 3 dimes, 3 nickels, 3 pennies	$_____	_____ ¢
1 five-dollar bill, 5 quarters, 7 nickels, 23 pennies	$_____	_____ ¢
9 one-dollar bills, 8 quarters, 12 dimes, 13 pennies	$_____	_____ ¢
4 ten-dollar bills, 1 five-dollar bill, 4 quarters, 1 nickel	$_____	_____ ¢

3 (a) 35¢ + [____] ¢ = $1   (b) 42¢ + $[____] = $1

(c) $0.71 + [____] ¢ = $1   (d) $0.92 + $[____] = $1

4 (a) $14.68 = [____] ¢   (b) 6,125¢ = $[____]

(c) $18.08 = [____] ¢   (d) 3,004¢ = $[____]

5 (a) How many nickels make $2.10?

(b) How many dimes make $8.50?

6 Alexus has 4 bills worth $25.00.
How many ten-dollar bills does he have?

7 Cooper has the same number of quarters and nickels.
He has $1.20 in quarters and nickels.
How many quarters does he have?

8 Ximena has 1 one-dollar bill, 5 quarters, and 2 dimes.
Ana has the same amount of money in nickels.
How many nickels does Ana have?

Challenge

9 Jamal has 30 quarters and dimes.
He has twice as much money in dimes as he has in quarters.
How much money does he have?

Basics

1 (a) Write the total amount of money.

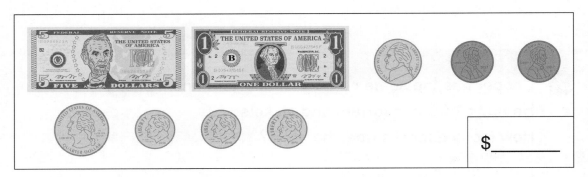

$ _____

(b) Adding $_____ in bills will make $9 in bills.

(c) Adding _____¢ in coins will make $1 in coins.

(d) $_____ more will make $10 in all.

2 (a) 35¢ + ☐ ¢ = $1   (b) $1.35 + ☐ ¢ = $2

  (c) $4.35 + ☐ ¢ = $5   (d) $9.35 + ☐ ¢ = $10

3 (a) $3 + $☐ = $9   (b) $0.45 + $☐ = $1

  (c) $3.45 + $☐ = $10

4 (a) $8 + $☐ = $9   (b) $0.52 + $☐ = $1

  (c) $8.52 + $☐ = $10

Practice

5

$3.32 E	$4.95 D	$6.50 N	$8.75
$2.46 O	$3.71 S	$5.10 R	$6.05 H
$9.70 N	$2.15 U	$9.29 E	$5.25 P
$1.14 E	$7.27 I	$1.40 N	$4.20 E
$5.63	$7.91 N	$8.00 D	$8.88

Joke: What has a hundred heads and a hundred tails?

Match an amount in the table above that makes $10 with an amount below.
Write the corresponding letter or space to find out.

$7.54	$3.50	$8.86	$1.25	$3.95	$7.85	$0.30	$5.05	$4.90	$5.80	$2.00

$4.37	$4.75	$6.68	$2.09	$8.60	$2.73	$0.71	$6.29	$1.12

6 Natalia buys a book that costs $4.46.
She pays for it with a ten-dollar bill.

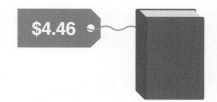

$4.46

(a) Find how much change she receives.

(b) Give two different sets of coins and bills she might receive as change.

Basics

1 (a) 65¢ + 60¢ = $ [ ]

35¢ [ ]

(b) 85¢ + 50¢ = $ [ ]

[ ] 50¢

(c) $3.75 + 40¢ = $ [ ]

25¢ [ ]

(d) $5.85 + 95¢ = $ [ ]

[ ] 5¢

2 (a) $29.35 —— **+ $12** ——→ $ [ ] —— **+ 65¢** ——→ $ [ ]

$29.35 + $12.65 = $ [ ]

(b) $17.40 —— **+ 60¢** ——→ $ [ ] —— **+ $11.25** ——→ $ [ ]

$17.40 + $11.85 = $ [ ]

(c) $36.45 —— **+ $15** ——→ $ [ ] —— **− 5¢** ——→ $ [ ]

$36.45 + $14.95 = $ [ ]

3 $47.48 ————→

$15.74 ————→

	4,	7	4	8	¢
+					¢
					¢

$47.48 + $15.74 = $ [ ]

Practice

 Add.

$73.45 + $0.65 **A**	$41.50 + $0.95 **B**	$43.55 + $25.80 **G**
$18.30 + $22.45 **N**	$28.85 + $46.70 **R**	$27.25 + $47.95 **E**
$31.65 + $31.65 **S**	$12.86 + $28.42 **C**	$11.93 + $12.84 **K**

What were the first U.S. bills called?

Write the letters that match the answers above to find out.

$69.35	$75.55	$75.20	$75.20	$40.75	$42.45	$74.10	$41.28	$24.77	$63.30

5 Makayla was paid $21.60 last week for helping a neighbor with yard work.
She was paid another $31.85 this week.
How much money did she earn altogether?

6 A shovel costs $37.99.
A rake costs $14.20 more than the shovel.
How much does the rake cost?

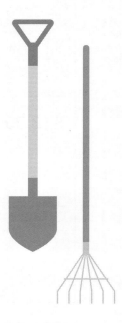

7 Raj bought a garden hose for $39.98, a sprinkler for $21.56, and a hose nozzle for $16.49.
How much money did he spend in all?

Basics

1 (a) $1 – 80¢ = $ [ ]

[ ]   80¢

(b) $1.55 – 80¢ = $ [ ]

55¢ [ ]

(c) $1 – 55¢ = $ [ ]

[ ]   55¢

(d) $5.15 – 55¢ = $ [ ]

$4.15 [ ]

2 (a) $20.35 ——**– $12**——→ $ [ ] ——**– 65¢**——→ $ [ ]

$20.35 – $12.65 = $ [ ]

(b) $17.40 ——**– 40¢**——→ $ [ ] ——**– $11.45**——→ $ [ ]

$17.40 – $11.85 = $ [ ]

(c) $36.45 ——**– $15**——→ $ [ ] ——**+ 5¢**——→ $ [ ]

$36.45 – $14.95 = $ [ ]

3 $42.46 ——————→   4, 2 4 6 ¢

$15.78 ——————→ – [ ][ ][ ][ ] ¢

[ ][ ][ ][ ] ¢

$42.46 – $15.78 = $ [ ]

Practice

 4 Subtract.

$43.45 – $0.65 **A**	$41.50 – $0.95 **N**	$43.55 – $25.80 **O**
$22.30 – $18.45 **I**	$46.70 – $28.85 **R**	$47.25 – $27.90 **G**
$41.65 – $22.70 **I**	$72.86 – $28.42 **O**	$51.23 – $12.84 **L**

The first known currency was created by King Alyattes in Lydia, now part of Turkey, in 600 BC. The first coin was stamped with a picture of what?

Write the letters that match the answers above to find out.

$42.80	$3.75	$17.85	$17.75	$42.80	$17.85	$3.85	$40.55	$19.35

$37.39	$38.39	$18.95	$44.44	$40.55

5 Liam has 1 twenty-dollar bill, 2 five-dollar bills, 6 quarters, 6 dimes, and 9 nickels.
Fernando has $32.85.
Who has more money, and how much more?

6 Olga spent $68.87 on a pair of shoes.
She paid with 4 twenty-dollar bills.
How much change did she receive?

7 Shivani spent $79.27 on a wheelbarrow and some topsoil.
If the topsoil cost $17.80, how much did the wheelbarrow cost?

Basics

1 A pair of pants costs $34.65.
A shirt costs $12.80 less than the pants.
A jacket costs $20.60 more than the shirt.
How much do the three items cost altogether?

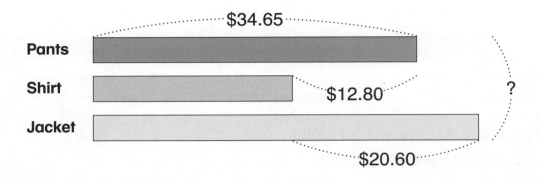

2 3 identical shirts cost $45.
A dress costs $5.20 more than a shirt.
How much do 5 shirts and 1 dress cost altogether?

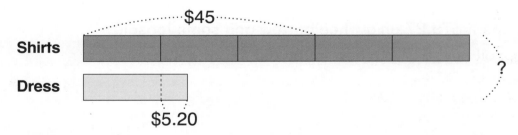

Practice

3 Isaac has $20.

He buys a computer mouse for $9.95 and a mouse pad for $2.80.

How much money does he have left?

4 A soccer ball costs $11.45.

It costs $4.98 less than a football.

How much do the two balls cost altogether?

5 A cap and 2 identical t-shirts cost $49.10.

A cap and 1 t-shirt cost $30.80.

How much does the cap cost?

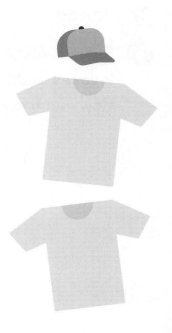

6 A set of speakers is usually priced at $29.50 and is on sale for $2 off.

A pair of headphones is usually priced at $19.15 and is on sale for $0.50 off.

Josef bought the speakers and headphones during the sale.

How much did he spend?

7 Nicole put $38 a month into savings for 9 months.

During that time, she took out $49.50 and another $15.75.

How much money did she save during the 9 months?

Challenge

8 5 vases and 2 placemats cost $51.

2 vases and 5 placemats cost $33.

How much does 1 placemat and 1 vase cost?

Check

1 (a) Write the amount of money in dollars and cents and in cents only.

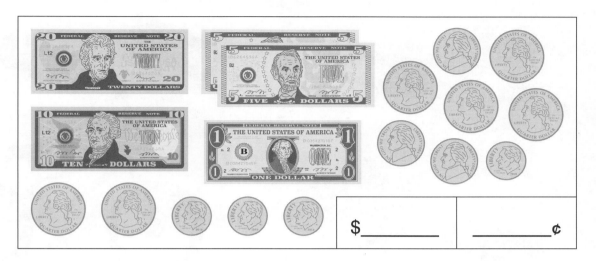

$ _____   _____ ¢

(b) How much more money will make $100?

2 Joshua has 26 quarters.

(a) If he trades in as many of his quarters as possible for dollar bills, how many quarters will he have left over?

(b) If he trades in all of his quarters for dimes, how many dimes will he have?

3 (a) $0.89 + $ [ ] = $1   (b) $ [ ] + $4.44 = $10

(c) $10.00 − $ [ ] = $6.21   (d) $20.00 − $ [ ] = $14.75

4

(a) How much more money does the helmet cost than the bat?

(b) Chapa bought the tennis racket and the 4-can pack of tennis balls.
She paid with 3 twenty-dollar bills and 3 ten-dollar bills.
How much change did she receive?

(c) Lucia bought the skateboard, the bike, and the helmet.
How much did she spend?

(d) Jody bought two items for $63.35.
Which two items did he buy?

(e) There are 3 tennis balls in each can and 4 cans cost $24.00.
At the same cost per can, what is the cost for 18 balls?

(f) Sharon bought the tennis racket and had it shipped.
She returned it and bought a different one that cost $10.85 more.
She paid $4.10 in shipping for each order.
How much did she spend altogether?

Challenge

5 Ada has some money to spend on some placemats.
If she were to buy 8 placemats, she would have $5 left.
If she were to buy 5 placemats, she would have $50 left.
How much money does she have?

6 John and Logan each have 6 bills that total $56.
John has 2 more five-dollar bills than Logan.
What bills does Logan have?

7 Each toy has a different cost.
The cost of some of them together is shown below.
What is the cost of each toy?

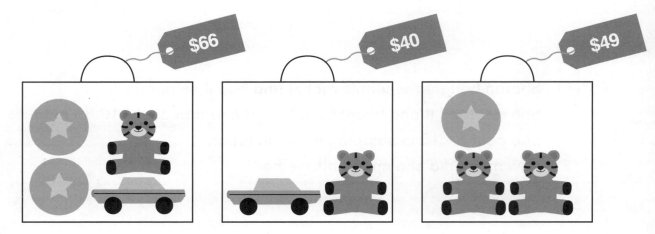

Check

1 What is the greatest 4-digit odd number that can be formed with the digits 4, 0, 9, and 2?

2 Divide 687 by 8.
What is the quotient and remainder?

3 Label the tick marks on this number line.
Each fraction should be in simplest form.

0   1

4 A jug can hold 2 L 250 mL of water.
A bottle can hold 1,340 mL less than the jug.
What is the total volume of the jug and the bottle in liters and milliliters?

5 The answer to each problem is the first number in the next problem.
Show what order to go through all the calculations to reach the end.

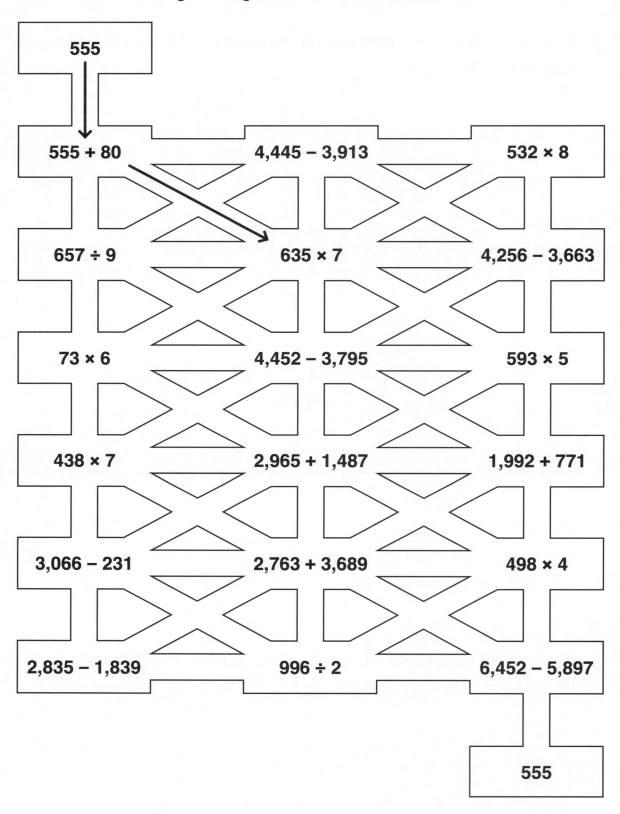

555

555 + 80   4,445 − 3,913   532 × 8

657 ÷ 9   635 × 7   4,256 − 3,663

73 × 6   4,452 − 3,795   593 × 5

438 × 7   2,965 + 1,487   1,992 + 771

3,066 − 231   2,763 + 3,689   498 × 4

2,835 − 1,839   996 ÷ 2   6,452 − 5,897

555

6 A square 8 units long was drawn and shaded as shown.

(a) What fraction of the square is shaded?
Give your answer in simplest form.

(b) What is the total area of the shaded parts?

7 How many eighths are in 2?

8 Light travels faster than sound.
To know how far away lightning strikes, you can count the seconds from when you see the lightning to when you hear the thunder.
Dividing this number by 5 will tell you about how many miles away the lightning struck.
Multiplying this number by 340 will tell you about how many meters you are from the lightning strike.

(a) If you count 66 seconds, about how many miles away did the lightning strike?

(b) If you count 9 seconds, about how many kilometers and meters away did the lightning strike?

9 The diameter of this circle is 14 cm.
One side of the triangle is 12 cm.

(a) What is the perimeter of the triangle?

(b) How many equal angles does the triangle have?

(c) How many angles of the triangle are smaller than a right angle?

(d) Two triangles like this one are put together along the 12-cm sides.
What kind of quadrilateral is formed and what is its perimeter?

10 On April 2, the sun was up for 12 hours, 10 minutes, and 11 seconds.
Yesterday, on April 1, the sun was up for 12 hours, 8 minutes, and 3 seconds.

(a) How much longer was the time between sunrise and sunset on April 2
than on April 1?

(b) In the next few weeks, the time between sunrise and sunset will increase
by the same amount.
How much daylight will there be on April 8?

11 A rectangular park is 120 m long and 60 m wide and has a path around it. How many whole laps does Lincoln have to run around it to have run at least a kilometer?

Challenge

12 Each hexagon below has 6 sides of equal length.
The perimeter of each hexagon is 37 cm.
What is the perimeter of the entire shape?

13 Kaiden has 5 times as much money as Paula.
After Kaiden spent $183, he had twice as much money as Paula.
How much money did Kaiden have at first?

Check

1 Find the missing numbers.

(a) $140 \times 6 = \boxed{} \times 3$

(b) $3 \times 72 = 6 \times \boxed{}$

(c) $245 \div 7 = 7 \times \boxed{}$

(d) $\boxed{} \times 2 = 108 \div 6$

2 If each lollipop costs a quarter, how many lollipops can Lucia buy with $4.80?

3 Rewrite any fraction in simplest form if it is not already in simplest form. Then list the fractions in order from least to greatest, using simplest form.

$\dfrac{8}{10}, \dfrac{3}{7}, \dfrac{5}{10}, \dfrac{6}{8}$

4 (a) Complete the table.

Circle	Radius	Diameter
A	_____ cm	3 m 4 cm
B	174 cm	_____ m _____ cm
C	1 m 85 cm	_____ cm
D	_____ m _____ cm	316 cm

(b) Circle _____ is the largest.

5 Poles A and B are both 1 m long.
$\frac{4}{9}$ of Pole A and $\frac{7}{9}$ of Pole B are painted green.
The rest of both poles are painted yellow.
Which pole has a longer length painted yellow, and how much longer is that length?

6 Laila left home at 10:45 a.m. and was gone for 4 hours and 25 minutes. What time did she arrive back home?

7 How much time passes in hours and minutes from 6:30 a.m. to 7:45 p.m.?

8 The information below shows how long it took some girls to swim 200 m.

Name	Ada	Cora	Holly	Mariam	Violet
Time	195 s	105 s	2 min 45 s	3 min 10 s	

(a) Violet's time was 15 s shorter than Mariam's time.
Add Violet's time to the chart.

(b) Complete a bar graph with the above information.

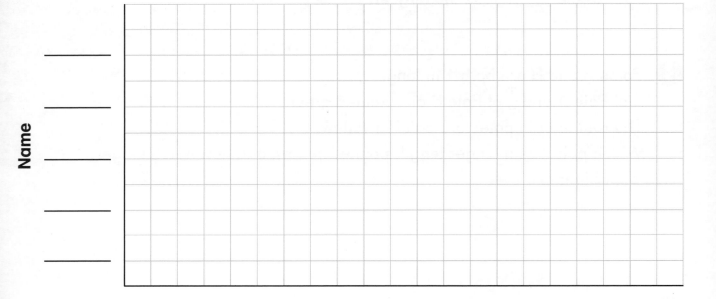

Time (seconds)

Name

(c) Put the times in order from shortest to longest.

(d) What is the difference between the longest time and the shortest time?

9 Find the area and perimeter of this figure.
All sides meet at right angles.

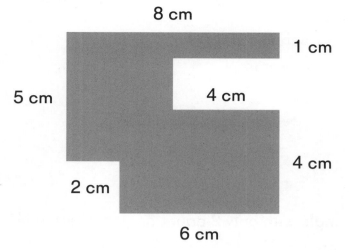

8 cm

1 cm

5 cm

4 cm

4 cm

2 cm

6 cm

10 All of the below circles are the same size.
Color the triangles that have at least two angles of the same size.

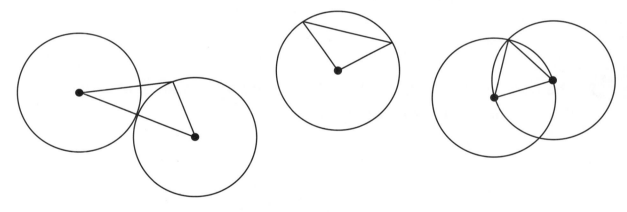

11 A farmer had 1,098 eggs.
He sold some of them.
He now has 145 cartons each with 6 eggs, and 4 additional eggs.
How many eggs has he sold so far?

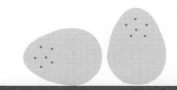

Challenge

12 200 pennies weigh 500 g.
Joseph has a collection of 700 pennies.
How much does his collection of pennies weigh?
Give your answer in kilograms and grams.

13 What is the perimeter of a triangle with only 2 equal angles where one side
is 3 cm and another side is 7 cm long?

14 The perimeter of a rectangle is 18 cm and the length of one side is 7 cm.
What is the area of the rectangle.

15 A rectangle covers an area of 11 unit squares.
The length of each side of the rectangle is a whole number of units.
What is the perimeter of the rectangle?